Springer Tracts in Advanced Robotics

Volume 29

Cristian Secchi · Stefano Stramigioli · Cesare Fantuzzi

Control of Interactive Robotic Interfaces

A Port-Hamiltonian Approach

With 86 Figures

Springer

Professor Bruno Siciliano, Dipartimento di Informatica e Sistemistica, Università degli Studi di Napoli Federico II, Via Claudio 21, 80125 Napoli, Italy, email: siciliano@unina.it

Professor Oussama Khatib, Robotics Laboratory, Department of Computer Science, Stanford University, Stanford, CA 94305-9010, USA, email: khatib@cs.stanford.edu

Professor Frans Groen, Department of Computer Science, Universiteit van Amsterdam, Kruislaan 403, 1098 SJ Amsterdam, The Netherlands, email: groen@science.uva.nl

Authors

Dr. Cristian Secchi
Prof. Dr. Cesare Fantuzzi

DISMI – University of Modena and Reggio Emilia
Morselli Building
Via G. Amendola 2
42100 Reggio Emilia
Italy

secchi.christian@unimore.it
cesare.fantuzzi@unimore.it

Prof. Dr. Stefano Stramigioli

Universiteit Twente
EL/CE
Postbus 217
7500 EAE Enschede
The Netherlands

S.Stramigioli@ieee.org

ISSN print edition: 1610-7438
ISSN electronic edition: 1610-742X

ISBN-10 3-540-49712-9 **Springer Berlin Heidelberg New York**
ISBN-13 978-3-540-49712-7 **Springer Berlin Heidelberg New York**

Library of Congress Control Number: 2006937681

Springer is a part of Springer Science+Business Media
springer.com

Printed in Germany

Typesetting: Digital data supplied by authors.
Data-conversion and production: PTP-Berlin Protago-TEX-Production GmbH, Germany (www.ptp-berlin.com)
Cover-Design: WMXDesign GmbH, Heidelberg
Printed on acid-free paper 89/3141/Yu - 5 4 3 2 1 0

STAR (Springer Tracts in Advanced Robotics) has been promoted under the auspices of EURON (European Robotics Research Network)

To Federica, Fiona and Stefania

Foreword

At the dawn of the new millennium, robotics is undergoing a major transformation in scope and dimension. From a largely dominant industrial focus, robotics is rapidly expanding into the challenges of unstructured environments. Interacting with, assisting, serving, and exploring with humans, the emerging robots will increasingly touch people and their lives.

The goal of the new series of *Springer Tracts in Advanced Robotics (STAR)* is to bring, in a timely fashion, the latest advances and developments in robotics on the basis of their significance and quality. It is our hope that the wider dissemination of research developments will stimulate more exchanges and collaborations among the research community and contribute to further advancement of this rapidly growing field.

The monograph written by Cristian Secchi, Stefano Stramigioli and Cesare Fantuzzi is focused on the role of energy in controlling physical systems. The port-Hamiltonian formalism is adopted which provides a framework for modeling physical systems based on the concepts of energy, interconnection and power ports describing the phenomena of energy storage, energy exchange and external interaction respectively. The potential of the work is to be found in the modelling and control of interactive robotic interfaces, such as haptic devices and telemanipulation systems.

Remarkably, the doctoral thesis at the basis of this monograph was a finalist for the Fifth EURON Georges Giralt PhD Award devoted to the best PhD thesis in Robotics in Europe. A fine addition to the series!

Naples, Italy,
September 2006

Bruno Siciliano
STAR Editor

Preface

The role of energy in modeling physical system is very well established and several modeling frameworks, such as Euler-Lagrange and Hamiltonian formalisms, are very well known since basic courses of physics. Despite its importance in modeling, the role of energy in controlling physical systems is not always recognized.

The port-Hamiltonian formalism provides a framework for modeling physical systems based on the concepts of energy, interconnection and power ports which model the phenomena of energy storage, energy exchange and external interaction respectively. Thus, within this framework the energetic properties of physical systems are very evident and it is possible to exploit them to build energy based controllers.

This work is based on the Ph.D. thesis of the first author which he defended at the University of Modena and Reggio Emilia (I) in 2004 and it deals with energy based control of interactive robotic interfaces. When two physical physical systems interact they exchange energy and in order to control the interaction in a sensible way, it is necessary to control this energy exchange. In this book, port-Hamiltonian framework is exploited both for modeling and controlling interactive robotic interfaces.

Starting from the port-Hamiltonian model, it is possible to identify the energetic properties that have to be controlled in order to achieve a desired interactive behavior and it is possible to build a port-Hamiltonian controller that properly regulates the robotic interface. Due to its generality, port-Hamiltonian formalism allows to deal also with complex interactive systems, such as haptic interfaces and telemanipulation systems, both linear and non linear, in a very intuitive way.

Many people contributed to the results presented in this book. In particular, the authors would like to thank Bernhard Maschke and Arjan van der Schaft whose remarks have always been very valuable. Furthermore, we would like to thank Alessandro Macchelli, Claudio Melchiorri and Nicola Diolaiti for the very pleasant and profitable collaborations and discussions.

A brief outline of the book is given in the following.

In Chapter 1, starting from the so-called behavioral approach for modeling, it is shown that the energy and the energetic interconnections, along which the internal power exchange takes place, and, the ports through which power is exchanged with the external world, are the essential ingredients to model the behavior of physical systems; the mathematical object of Dirac structure is introduced and both implicit and explicit port-Hamiltonian systems are introduced.

In Chapter 2 energy is exploited for control purpose and it is shown how to exploit the energetic properties of port-Hamiltonian systems to build energy-based controllers that allow to solve the regulation problem for physical systems. The basics of passivity theory are presented and the link between stability of a certain configuration and the shape of the energy of the physical system is illustrated. The energy shaping regulation technique for port-Hamiltonian systems is presented both from an energy balancing and an interconnection an damping assignment perspective. A control strategy that embeds variable structure techniques in energy based control is illustrated. The resulting control scheme allows to enhance robustness and performances in regulation tasks and it is the result of a collaboration with Alessandro Macchelli and Claudio Melchiorri from the University of Bologna.

In Chapter 3 the problem of controlling interactive robotic interfaces is introduced. An energetic analysis of interaction is provided and an intrinsically passive control strategy for interactive systems is presented; port-Hamiltonian controllers are used to shape the energetic properties of the robotic interface and, therefore, to achieve the desired kind of contact behavior. A port-Hamiltonian impedance controller (also called IPC) is introduced and some new developments that allow to deal also with defective anthropomorphic robots and with complex robotic interfaces as robotic hands are illustrated. Furthermore, an energetic model of a generic haptic interface is presented and the port-Hamiltonian formalism is used for building an intrinsically passive control scheme for haptic interfaces which allows a stable interaction with generic, both linear and nonlinear, virtual environments. Finally, exploiting the port-Hamiltonian formalism, some typical problems related to haptic interfaces as that of delayed virtual environments and of force scaling are analyzed and solved.

In Chapter 4 bilateral telemanipulation systems, that allows to interact with remote environment, are introduced and an energy based analysis is provided. Port-Hamiltonian formalism together with scattering theory are used to achieve an intrinsically passive port-Hamiltonian based bilateral telemanipulation scheme which exhibits a stable behavior both in case of contact with the remote environment and of free motion, independently of any delay. It is shown how to extend the scheme in order to passively deal with the discrete nature of the controllers and with variable delay packet switching communication channels (e.g. Internet). Finally, a passivity preserving interpolation algorithm is introduced in order to improve performances of the telemanipulation scheme in case of loss of packets in the communication.

In Chapter 5 the problem of transparency in bilateral telemanipulation is illustrated. A framework, based on the behavioral approach, for the evaluation of transparency is presented and used to evaluate transparency of port-Hamiltonian based intrinsically passive bilateral telemanipulation schemes. Extended port-Hamiltonian systems are introduced to allow a passivity preserving variation of physical parameters characterizing the port-Hamiltonian controllers used to control master and slave robots. These results are the outcome of a collaboration with Nicola Diolaiti from the University of Bologna. A novel scheme that allows to increase transparency in port-Hamiltonian based bilateral telemanipulation is proposed using the variable parameters IPC.

Finally, in Appendix A, some background on the mathematical tools used in the book is provided.

Reggio Emilia (Italy),
Enschede (The Netherlands),
August 2006

Cristian Secchi
Stefano Stramigioli
Cesare Fantuzzi

Contents

1 Physical Modeling and Port-Hamiltonian Systems 1
- 1.1 Introduction 1
- 1.2 The Behavioral Paradigm for Modeling Dynamical Systems 2
  - 1.2.1 Universe, Behavior and Behavioral Equations 2
  - 1.2.2 Manifest and Latent Variables 3
  - 1.2.3 Dynamical Systems 4
  - 1.2.4 Inputs and Outputs 8
- 1.3 Physical Modeling 10
- 1.4 Implicit Port-Hamiltonian Systems 14
  - 1.4.1 A Coordinate Based Representation 18
- 1.5 Geometric Scattering 25
- 1.6 Conclusions 30

2 Control of Port-Hamiltonian Systems 33
- 2.1 Introduction 33
- 2.2 Basic Concepts of Passivity Theory 34
  - 2.2.1 Definitions and Properties 34
  - 2.2.2 Output Feedback Stabilization of Passive Systems 38
  - 2.2.3 Port-Hamiltonian Systems and Passivity 40
- 2.3 Energy Shaping of Port-Hamiltonian Systems 44
  - 2.3.1 Stabilization by Energy Balancing 45
  - 2.3.2 The Control as Interconnection Paradigm 51
  - 2.3.3 Energy Shaping as Control by Interconnection 55
- 2.4 Interconnection and Damping Assignment Passivity Based Control 60
- 2.5 A Variable Structure Approach to Energy-Based Control 64
- 2.6 Conclusions 74

3 A Port-Hamiltonian Approach to the Control of Interaction 77
- 3.1 Introduction 77
- 3.2 Intrinsically Passive Control of Interaction 78

3.3 Intrinsically Passive Control of Robotic Systems 80
3.3.1 IPC for Anthropomorphic Robotic Arms 81
3.3.2 IPC for Grasping 86
3.4 Interaction with Virtual Environments: Haptic Interfaces 88
3.4.1 Sampled Port-Hamiltonian Systems 91
3.4.2 Energy Consistent Sampled Passivity 97
3.4.3 Passive Coupled Behavior 100
3.4.4 Dealing with Quantization Errors 101
3.5 Delayed Virtual Environments 104
3.5.1 The Effect of Delayed Output 104
3.5.2 Passive Discretization of Port-Hamiltonian Systems in Scattering Representation 108
3.6 Force Scaling in Port-Hamiltonian Based Haptic Interfaces.... 110
3.6.1 Power Scaling in Port-Hamiltonian Based Haptic Interfaces 112
3.6.2 Variable Scaling 114
3.7 Simulations 117
3.8 Conclusions 124

4 Port-Hamiltonian Based Bilateral Telemanipulation 127
4.1 Introduction 127
4.2 Port-Hamiltonian Based Bilateral Telemanipulation 129
4.2.1 An Energetic Analysis of a Bilateral Telemanipulation System 129
4.2.2 Passive Control of Interaction 131
4.2.3 Passive Communication Channel 132
4.2.4 The Intrinsically Passive Telemanipulation Scheme 137
4.3 Complex Telemanipulation Systems: Telegrasping 142
4.4 A Digital Scheme for Intrinsically Passive Telemanipulation ... 147
4.5 Improving Performances in Intrinsically Passive Digital Telemanipulation 155
4.6 Conclusions 162

5 Transparency in Port-Hamiltonian Based Telemanipulation 165
5.1 Introduction 165
5.2 A Model for the Contact Impedance 167
5.3 A Behavioral Framework for Evaluating Transparency 168
5.3.1 Analysis of the Port Behavior 168
5.3.2 Transparency in Telemanipulation 171
5.3.3 Transparent Telemanipulation as a Behavioral Control Problem 172

5.4 Transparency in Port-Hamiltonian Based Telemanipulation . . . 173
5.4.1 Tuning the IPC . . . 174
5.4.2 Transparency Analysis of a Scattering Based Packet Switching Communication Channel . . . 175
5.4.3 Simulations . . . 182
5.5 A Passivity Preserving Tuning of Port-Hamiltonian Systems . . 184
5.5.1 Parameters Associated to Energy Storage . . . 187
5.5.2 Parameters Associated to Energy Dissipation and Interconnection . . . 194
5.5.3 Simulations . . . 194
5.6 A Scheme for Transparent Port-Hamiltonian Based Telemanipulation . . . 196
5.7 Conclusions . . . 198

A Mathematical Background . . . 201
A.1 Manifolds and Vector Bundles . . . 201
A.2 Tensors . . . 206
A.3 Lie Groups and Rigid Motions . . . 207
A.3.1 An Example: The Special Euclidean Group SE(3) 208

References . . . 211

Index . . . 231

List of Tables

1.1 Efforts and flows for various physical domains 13

2.1 Parameters of the considered manipulator. 70

List of Figures

1.1 A simple resistive network 4
1.2 An LC circuit 6
1.3 A mass-spring system 11
1.4 Oscillator as the interconnection of mass and spring 12
1.5 Network model of a physical system 14
1.6 Schematic representation of a DC motor 23
1.7 Schematic representation of a n-DOF anthropomorphic robot . . 24
1.8 Scattering power decomposition 30

2.1 The physical equivalent of the oscillator controlled by damping injection 44
2.2 Series RLC circuit 49
2.3 Parallel RLC circuit 50
2.4 Control as interconnection 52
2.5 Power preserving interconnection of port-Hamiltonian systems . 53
2.6 Energy storing element behavior. 65
2.7 Energy and force of non saturated (regular) and saturated (bold) spring. 66
2.8 Behavior of the proposed controller. 69
2.9 A planar 2-DOF manipulator 70
2.10 Simulation results with PD+g(q). 71
2.11 Simulation results: errors, torques and velocities 72
2.12 Simulation results: energy. 73
2.13 Simulation results with PD + g(q) and partial knowledge of mass parameters: errors. 73
2.14 Simulation results with partial knowledge of mass parameters: errors. 74
2.15 Simulation results with partial knowledge of mass parameters and saturation. 75

3.1 The general form of an intrinsically passive control scheme 80

3.2 The IPC for anthropomorphic robots 81
3.3 The 1-DOF variable length spring. 83
3.4 A two DOF defective robot 85
3.5 Rotational velocity $\omega_v^{b,b}$ and displacement (dashed) at the the variable length port of the spatial spring. 86
3.6 Desired motion of the planar robot. 86
3.7 The grasping system 87
3.8 Energetic representation of a haptic display 89
3.9 Active behavior of a virtual spring 90
3.10 The passivity-preserving energetic behavior of a discrete port-Hamiltonian system 94
3.11 The various possible connections for the state jump 95
3.12 The energy leap strategy 96
3.13 Energy Booking for a Forward Jump 97
3.14 The interconnection of discrete and continuous port-Hamiltonian systems 98
3.15 The sample & hold plus dissipation 102
3.16 Codification of power variables 106
3.17 The final scheme 109
3.18 Energetic representation of a haptic display 112
3.19 Energetic behavior in the ideal and in the noisy case. Continuous Energy (solid) and Discrete Energy (dashed) 118
3.20 Compensations. Continuous (solid) and discrete (dashed) energy 119
3.21 Position of the mass 119
3.22 Position of the mass 120
3.23 Energetic behavior of the spring-mass system 120
3.24 Position of the mass in case of a computational delay $\delta = 0.05$ on the output of the spring 121
3.25 Position of the mass in case of computational delay $\delta = 0.05sec$: scattering representation 121
3.26 Virtual Wall 122
3.27 An abrupt change of stiffness 123
3.28 A gradual change of stiffness 123
3.29 The proposed scaling strategy 124

4.1 A schematic representation of a telemanipulation scheme 130
4.2 A detailed representation of a generic bilateral telemanipulation scheme 130
4.3 A possible implementation of a communication channel for force reflection 132
4.4 The scattering based communication channel 135
4.5 The port-Hamiltonian Telemanipulation Scheme 137
4.6 Teleoperator with anthropomorphic robotic arms 140
4.7 Free motion with and without impedance matching. Positions of the master (solid) and of the slave (dashed) 141

4.8 Interaction task. Positions of the master (solid) and of the slave (dashed) . . . 141
4.9 Position of the fingers . . . 142
4.10 The power at master side . . . 143
4.11 The power at slave side . . . 144
4.12 Position of the fingers . . . 145
4.13 The power at master side . . . 146
4.14 The power at slave side . . . 147
4.15 Position of the fingers in a grasp . . . 148
4.16 The Passive Sampled Data Telemanipulation Scheme . . . 149
4.17 The scattering based communication channel . . . 150
4.18 Free motion with different sample times. Position of the master (continuous) and of the slave (dashed) . . . 154
4.19 Variable delay and loss of packets . . . 155
4.20 Position of the master (continuous) ad of the slave (dashed) . . . 155
4.21 The Communication Channel . . . 157
4.22 Positions of Master and Slave . . . 161
4.23 Positions of Master and Slave in case of loss of packets . . . 161
4.24 Positions of Master and Slave in case of loss of packets and interpolation algorithm enabled . . . 162

5.1 A generic teleoperator . . . 171
5.2 The role of IPC . . . 174
5.3 $\varepsilon(t)$: effect of controllers parameters . . . 183
5.4 Effect on transparency of loss of packets in the communication . 184
5.5 Effect on transparency of loss of packets in the communication when interpolation is enabled . . . 185
5.6 Effect on transparency of variable delay in the communication . 186
5.7 A 1-DOF variable parameters port-Hamiltonian impedance controller . . . 195
5.8 Stiffness tracking and spring elongation . . . 197
5.9 Transparency optimized intrinsically passive telemanipulation scheme . . . 198

A.1 Compatible charts . . . 202

1

Physical Modeling and Port-Hamiltonian Systems

1.1 Introduction

Interaction between physical systems is determined by an exchange of energy and, therefore, a first step towards the control of interaction is to explicitly model the energetic properties of physical systems.

First of all, *what is a mathematical model in general?* Let a phenomenon to model be given. The phenomenon produces *outcomes* (*events*) and the model has to represent the outcomes that can occur. For example, the principles of thermodynamics limit the amount of heat that can be transformed into mechanical work and relativity theory tells that a body cannot travel faster than the speed of light. In general, a mathematical model selects a certain subset from a universe of possibilities; this subset contains the outcomes that are declared possible by the model, the *behavior* of the model. Equations are a very suitable tool to select a certain subset from a universe of possibilities: a certain occurrence of the phenomenon is forbidden if a certain constitutive equation is not satisfied. The behavior of the model, thus, is given by those occurrences that satisfy a certain set of constitutive equations, called *behavioral equations.* A model has to represent certain variables of main interest which are called *manifest variables.* On the other hand, very often a mathematical model involves the use of other auxiliary variables, called *latent variables* to model the behavior of the phenomenon.

Behavior, behavioral equations, manifest and latent variables are the three main ingredients of a modeling language called *behavioral approach* that has been introduced in [330, 243]. This approach allows to describe phenomena in a very general way without any necessary a priori distinction between inputs and outputs (which is not always clear: when modeling a linear resistor, the voltage has to be considered as an input and the current as an output or vice versa?) which is instead very common when using standard modeling techniques as, for example, transfer functions or transfer matrixes.

Because of its generality, the behavioral approach is a suitable framework for describing physical systems; these systems have a very particular structure that yields a lot of information that can be used in the modeling process. In particular it is possible to define energy and to model a system by means of variables that are related to energy storage and by the way energy is exchanged among the constitutive parts of the system. Furthermore it is possible to define the concept of power port, namely the medium through which the system exchanges energy with the rest of the world.

In this chapter we will show how it is possible to embed the physical nature of the system into the behavioral framework. We will first provide an introduction, based on [330, 243, 295], of the behavioral approach for modeling a dynamical system. Once we have a general framework for modeling any phenomenon, we will focus on physical system and we will introduce the concepts of energy, energy variable and power port that are fundamental for describing the phenomena of energy storage and of energy exchange. We will then present the *port-Hamiltonian formalism* that has been introduced in [189, 323, 322, 318, 319] and that allows to model any physical system by taking explicitly into account its energetic behavior and its internal interconnection structure and we will show how physics naturally determines the kind of behavior of any physical system. Finally, we will present the scattering representation of power ports and the concept of scattering variables, based on the treatment reported in [299], that will be very useful when dealing with complex interacting systems with delay.

1.2 The Behavioral Paradigm for Modeling Dynamical Systems

1.2.1 Universe, Behavior and Behavioral Equations

Each phenomenon produces *outcomes* (events) which take value in a set $\mathcal{U}$, called *Universe*. Usually not all the elements of the universe are possible outcomes for the given phenomenon; to define a model means to determine whether certain outcomes are possible or not. A model, therefore, selects a certain subset $\mathcal{B}$, which represents the possible outcomes, of the universe $\mathcal{U}$; $\mathcal{B}$ is called the *behavior* of the mathematical model[243].

Definition 1.1 (Mathematical Model). *A* mathematical model *is a pair* $(\mathcal{U}, \mathcal{B})$. $\mathcal{U}$ *is a set, called the* universe, *whose elements are called* outcomes *and* $\mathcal{B}$ *is a subset of* $\mathcal{U}$, *called the* behavior.

Example 1.2 (Willems 1991). It is well known that H_2O can appear, depending on the temperature, as ice, liquid water or steam. The universe is given, with the temperature expressed in Celsius, by the set

$$\mathcal{U} = \{ice, water, steam\} \times [-273, \infty)$$

and the behavior by the subset

$$\mathcal{B} = ((\{ice\} \times [-273, 0]) \cup (\{water\} \times [0, 100]) \cup (\{steam\} \times [100, \infty)))$$

A certain behavior can be selected from a universe in several ways: by simple enumeration (as in Example 1.2), by the fact of having certain properties, etc. A very effective way to describe a phenomenon is by means of equations: an outcome of the universe is part of the behavior if and only if it satisfies certain constitutive equations.

Definition 1.3 (Behavioral equations). *Let $\mathcal{U}$ be a universe, $\mathcal{E}$ a set and $f_1, f_2 : \mathcal{U} \to \mathcal{E}$ two maps. The mathematical model $(\mathcal{U}, \mathcal{B})$ with $\mathcal{B} = \{u \in \mathcal{U} \mid f_1(u) = f_2(u)\}$ is said to be described by* behavioral equations *and is denoted by $(\mathcal{U}, \mathcal{E}, f_1, f_2)$. The set $\mathcal{E}$ is called the* equating space. *The quadruple $(\mathcal{U}, \mathcal{E}, f_1, f_2)$ is also called a* behavioral equation representation *of $(\mathcal{U}, \mathcal{B})$.*

Example 1.4. Consider a linear resistor. It is well known that the relation between the current I flowing through the wire and the voltage V across the terminals is determined by Ohm's law:

$$V = RI$$

The system can be modeled by considering as universe $\mathcal{U} = \mathbb{R}^2$, as equating space $\mathcal{E} = \mathbb{R}$. Consider:

$$f_1 : (V, I) \to V \qquad f_2 : (V, I) \to RI$$

the behavior of the resistor can be given as

$$\mathcal{B} = \{(V, I) \in \mathcal{U} \mid f_1(V, I) = f_2(V, I)\}$$

which means nothing else that

$$\mathcal{B} = \{(V, I) \in \mathbb{R}^2 \mid V = RI\}$$

1.2.2 Manifest and Latent Variables

Given a certain phenomenon to describe, a mathematical model can be built in two steps: first identify the variables to model and their domain of existence, the universe $\mathcal{U}$, second identify all the possible outcomes, the behavior $\mathcal{B} \subset \mathcal{U}$. The behavior can be specified by requiring that some behavioral equations are satisfied. In order to express the model in a convenient way, it can be necessary to introduce some auxiliary variables in addition to those that have to be modeled. These extra variables are called *latent variables* and live in a certain domain $\mathcal{U}_l$ while the variables that have to be explicitly modeled are called *manifest variables* and live in the universe $\mathcal{U}$.

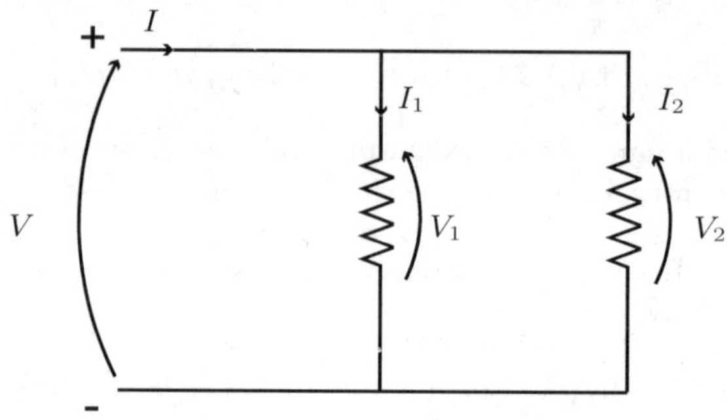

Fig. 1.1. A simple resistive network

Example 1.5. Consider the simple resistive network represented in Fig. 1.1 and assume that the behaviors of the current I and of the voltage V at the external port have to be modeled. The physical principles to use to model the port behavior are Kirchhoff's laws and Ohm's law. V and I are manifest variables while I_1, I_2, V_1 and V_2 are latent variables. The behavioral equations are:

$$\begin{aligned} R_1 I_1 &= V_1 \qquad & V_1 &= V_2 = V \\ R_2 I_2 &= V_2 \qquad & I_1 + I_2 &= I \end{aligned} \tag{1.1}$$

It is now possible to give the following:

Definition 1.6 (Mathematical model with latent variables). *A mathematical model with latent variables is defined as a triple $(\mathcal{U}, \mathcal{U}_l, \mathcal{B}_f)$ where $\mathcal{U}$ is the universe of manifest variables, $\mathcal{U}_l$ is the universe of latent variables and $\mathcal{B}_f \subset \mathcal{U} \times \mathcal{U}_l$ is the full behavior. The manifest mathematical model $(\mathcal{U}, \mathcal{B})$ is then defined by:*

$$\mathcal{B} = \{u \in \mathcal{U} \mid \exists l \in \mathcal{U}_l \text{ such that } (u, l) \in \mathcal{B}_f\} \tag{1.2}$$

$\mathcal{B}$ is called the manifest behavior or, simply, behavior and the triple $(\mathcal{U}, \mathcal{U}_l, \mathcal{B}_f)$ is called a latent variable representation of the manifest mathematical model $(\mathcal{U}, \mathcal{B})$.

1.2.3 Dynamical Systems

Dynamical systems are characterized by a time evolution and are of crucial importance in control theory. It is possible to describe these systems in the context of Def. 1.1: the variables of interests are function of time and, therefore, the universe is a function space.

Definition 1.7 (Dynamical system). *A dynamical system is defined by a triple $(\mathcal{T}, \mathcal{W}, \mathcal{B})$ where $\mathcal{T}$ is the time axis, $\mathcal{W}$ is a set called* signal space *and $\mathcal{B} \subset \mathcal{W}^{\mathcal{T}}$ is called the behavior. Here, $\mathcal{W}^{\mathcal{T}}$ indicates the set of maps from $\mathcal{T}$ to $\mathcal{W}$.*

The time axis represents the ordered set of time instants of interest, the signal space $\mathcal{W}$ represents the set where the outcomes of the system evolve and $\mathcal{W}^{\mathcal{T}}$ represents the universe, namely the set of all trajectories taking value in the signal space. Analogously to mathematical models, the behavior $\mathcal{B}$ is nothing else than a subset of the universe which selects a family of time trajectories that are compatible with the system.

Definition 1.8 (Time invariance). *A dynamical system $\Sigma = (\mathcal{T}, \mathcal{W}, \mathcal{B})$ is said to be* time invariant *if $\sigma^t \mathcal{B} = \mathcal{B}$ for all $t \in \mathcal{T}$, where σ^t is the* t-shift operator *defined as:*

$$(\sigma^t f)(\tau) = f(\tau + t) \tag{1.3}$$

The time invariance property says that the behavior of the dynamical system does not depend explicitly by the time, as well known from system theory [128, 284].

In this chapter we aim at modeling physical systems and, therefore, the behavioral framework will be illustrated for continuous dynamical systems and thus $\mathcal{T} = \mathbb{R}$. Furthermore we will assume that the signal space is a differentiable manifold (this is not restrictive for the modeling purposes and it is an essential technical assumption for many definitions) and we will assume that the dynamical systems under consideration are time invariant. The results illustrated in this section can be easily extended to discrete time dynamical systems, as shown in [330].

It is possible to describe the behavior of a dynamical system by means of differential equations and this can be formalized by using the concept of jet space [221]. Consider a differentiable manifold $\mathcal{G}$ and take as time axis $\mathbb{R}$. Consider a sufficiently smooth map $g(\cdot) \in \mathcal{G}^{\mathbb{R}}$, i.e. such that its nth derivative with respect to t is defined. Let $\mathcal{G}_i$ be the set of all possible instantaneous i^{th} time derivative for any possible $g(\cdot) \in \mathcal{G}^{\mathbb{R}}$. It is then possible to define the set $\mathcal{G}^{(n)} = \mathcal{G} \times \mathcal{G}_1 \times \cdots \times \mathcal{G}_n$; points in $\mathcal{G}^{(n)}$ are indicated with $g^{(n)}$. The induced map $g^{(n)}(\cdot) = pr^{(n)} g(t)$, called the nth *prolongation of* $g(\cdot)$, is given by:

$$g^{(n)}(t) = (g(t), g_1(t), \cdots, g_n(t)) \tag{1.4}$$

where

$$g_i(t) = \left.\frac{d^i g}{dt^i}\right|_t \tag{1.5}$$

Thus $pr^{(n)} g(\cdot)$ is a map from the time axis $\mathbb{R}$ to the space $\mathcal{G}^{(n)}$.
Consider a dynamical system $\Sigma = (\mathbb{R}, \mathcal{W}, \mathcal{B})$:

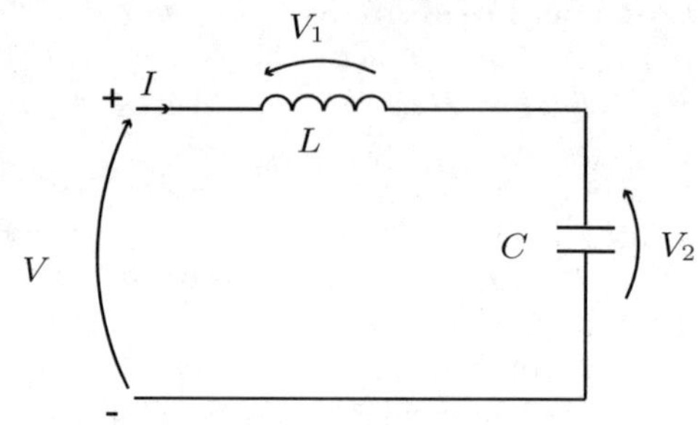

Fig. 1.2. An LC circuit

Definition 1.9 (Jet space). *The nth order jet space of $\mathbb{R} \times \mathcal{W}$ is defined as the space $\mathbb{R} \times \mathcal{W}^{(n)}$ and it will be indicated as $J^n(\mathbb{R} \times \mathcal{W})$.*

An element in the signal space $\mathcal{W}$ can be considered as the value taken by a map $w \in \mathcal{W}^{\mathbb{R}}$ in a certain point $t \in \mathbb{R}$. Thus, an element of $J^n(\mathbb{R} \times \mathcal{W})$ is given by an element of the signal space $w(t) \in \mathcal{W}$ together with its derivatives calculated in t up to order n.

Definition 1.10 (Dynamical behavioral equations). *Let $\mathcal{W}$ be a differentiable manifold which represents the signal space, $\mathbb{R}$ be the time axis, n be a nonnegative integer(the* order *of the differential equation) and $\mathcal{E}$ be the* equating space. *Consider two maps:*

$$f_1, f_2 : J^n(\mathbb{R} \times \mathcal{W}) \to \mathcal{E} \tag{1.6}$$

The dynamical system $\Sigma = (\mathbb{R}, \mathcal{W}, \mathcal{B})$ with

$$\mathcal{B} = \{w : \mathbb{R} \to \mathcal{W} \mid f_1(w(t), \frac{dw}{dt}, \cdots, \frac{d^n w}{dt^n}) = f_2(w(t), \frac{dw}{dt}, \cdots, \frac{d^n w}{dt^n})\} \tag{1.7}$$

is said to be described by differential equations and is denoted by $(\mathbb{R}, \mathcal{W}, \mathcal{E}, f_1, f_2)$

Example 1.11. Consider the simple LC circuit represented in Fig. 1.2. Suppose that a model of the behavior of the port variables V and I is needed. In this case the time axis is $\mathbb{R}$ and the signal space is $\mathbb{R}^2$. From the constitutive equations of the inductor and of the capacitor and from Kirchhoff's laws it follows that:

$$\begin{cases} V = V_1 + V_2 \\ L\dfrac{dI}{dt} = V_1 \\ C\dfrac{dV_2}{dt} = I \end{cases} \tag{1.8}$$

Consider now the following maps:

$$
\begin{aligned}
f_1 &: J(\mathbb{R} \times \mathbb{R}^2) \to \mathbb{R} \; f_1(pr^{(2)}(V, I)) = L\frac{d^2 I}{dt^2} \\
f_2 &: J(\mathbb{R} \times \mathbb{R}^2) \to \mathbb{R} \; f_2(pr^{(2)}(V, I)) = \frac{dV}{dt} - \frac{1}{C}I
\end{aligned}
\tag{1.9}
$$

By straightforward calculations, it follows that the behavior of the system is governed by the differential equation:

$$f_1(pr^{(2)}(V, I)) = f_2(pr^{(2)}(V, I)) \tag{1.10}$$

namely by:

$$L\frac{d^2 V}{dt^2} = \frac{dV}{dt} - \frac{1}{C}I \tag{1.11}$$

It is straightforward to generalize the notion of latent and manifest variables for dynamical systems.

Definition 1.12 (Dynamical system with latent variables). *A dynamical system with latent variables is defined as* $\Sigma_L = (\mathbb{R}, \mathcal{W}, \mathcal{L}, \mathcal{B}_f)$ *where* $\mathbb{R}$ *is the time axis,* $\mathcal{W}$ *is the manifest signal space,* $\mathcal{L}$ *is the latent variables space and* $\mathcal{B}_f \subset (\mathcal{W} \times \mathcal{L})^{\mathbb{R}}$ *is the full behavior. It defines a latent variable representation of the manifest dynamical system* $\Sigma = (\mathbb{R}, \mathcal{W}, \mathcal{B})$ *with the following manifest behavior:*

$$\mathcal{B} = \{w : \mathbb{R} \to \mathcal{W} \mid \exists l : \mathbb{R} \to \mathcal{L} \; such\ that (w, l) \in \mathcal{B}_f\} \tag{1.12}$$

Latent variables can appear in modeling dynamical systems for various reasons. First, they can be used simply for mathematical reasons, in order to obtain a particular form of the behavioral equations. Second, they can be used to express internal laws governing the behavior of the system. A few examples are internal current and voltages to express the behavior of an electrical circuit at external ports and momentum in classical Hamiltonian mechanics to express the evolution of position. Furthermore, latent variables appear when modeling a system which is composed by the interconnection of several subsystems; in this case latent variables will express the interconnection constraints. A very important class of latent variables in system theory are state variables whose role is to represent the internal memory of the system.

Definition 1.13 (State space dynamical system). *A* state space dynamical system *is a dynamical system with latent variables* $\Sigma_L = (\mathbb{R}, \mathcal{W}, \mathcal{X}, \mathcal{B}_f)$ *in which the full behavior satisfies the* axiom of state. *The axiom of state requires what follows:*
Given $(w_1, x_1), (w_2, x_2) \in \mathcal{B}_f$ *and* $t \in \mathbb{R}$, *if* $x_1(t) = x_2(t)$ *then* $(w, x) \in \mathcal{B}_f$, *where:*

$$
(w(\tau), x(\tau)) = \begin{cases} (w_1(\tau), x_1(\tau)) \; for\, \tau < t \\ (w_2(\tau), x_2(\tau)) \; for\, \tau \geq t \end{cases}
$$

The axiom of state requires that any possible trajectory if the full behavior $\mathcal{B}_f$ ending in a particular state can be concatenated with any other trajectory starting from that state. In other words, it asserts that once that the initial state is known, there is enough information to determine the future behavior and there is no need of further information of the past trajectory. The axiom of state formalizes a concept very well known since basic system theory courses: *the state of a dynamical system contains sufficient information about the past in order to determine the future behavior.*

A question that arises very naturally at this point is: *How can state space dynamical systems be represented by differential behavioral equations?*
The following theorem holds:

Theorem 1.14 (Willems 1991). *Let $\Sigma_L = (\mathbb{R}, \mathcal{W}, \mathcal{X}, \mathcal{B}_f)$ be a dynamical system with latent variables and let $\mathcal{X}$ be a differentiable manifold and $T\mathcal{X}$ its tangent bundle. Consider two maps $f_1, f_2 : T\mathcal{X} \times \mathcal{W} \to \mathcal{E}$ where $\mathcal{E}$ is the equating space . The differential equation:*

$$f_1((x, \dot{x}), w) = f_2((x, \dot{x}), w)$$

defines a state space system $(\mathbb{R}, \mathcal{W}, \mathcal{X}, \mathcal{B}_f)$ with

$\mathcal{B}_f = \{(w, x) : \mathbb{R} \to \mathcal{W} \times \mathcal{X} \mid x$ *is absolutely continuous and* $f_1(x(t), \dot{x}(t), w(t)) = f_2(x(t), \dot{x}(t), w(t))$ *for almost all* $t \in \mathbb{R}\}$

Theorem 1.14 shows that the fundamental condition for a state space dynamical system is that the behavioral differential equations are of first order in x and of zero order in w. The theorem does NOT state that the behavioral differential equations must be necessarily explicit, i.e. of the form:

$$\dot{x} = f(x, w) \tag{1.13}$$

The representation given for state space dynamical system is therefore able to represent both explicit and implicit systems.

1.2.4 Inputs and Outputs

So far no distinction has been made between inputs and outputs. This is not a weak point of the behavioral approach, as it could be stated at a first glance, rather its strength. Inputs and outputs should be *deduced from* the mathematical model of a certain phenomenon and not *imposed on* the model. In fact, for example, in physical systems it is not always clear which variable should be chosen as input and which as output. Considering an electrical resistor, it is not clear whether voltage has to be taken as input and current as an output or vice versa. Furthermore, the choice of input and output could depend on the operating point of the system. Finally, the choice of inputs and outputs could also depend on the specific task the system has to accomplish. It is then clear that in order to get a general mathematical model, it is better to use the concept of outcome rather than specify *a priori* inputs and outputs. However it is important to be able to include inputs and outputs in the behavioral framework because of several reasons (e.g. for control purposes) it can be necessary to act on some inputs in order to get a desired output.

In the following we explain how it is possible to give an input ut representation of a dynamical system:

Definition 1.15 (I/O dynamical system). *Consider a behavior $\mathcal{B}$ a signal space $\mathcal{W} = \mathcal{U} \times \mathcal{Y}$. An* I/O dynamical system *is defined as the quadruple $\Sigma_{I/O} = (\mathbb{R}, \mathcal{U}, \mathcal{Y}, \mathcal{B})$, where $\mathbb{R}$ represents the time axis, $\mathcal{U}$ the* input signal space, *$\mathcal{Y}$ the* output signal space *and $\mathcal{B} \subset (\mathcal{U} \times \mathcal{Y})^{\mathbb{R}}$ the behavior. The following properties have to be satisfied:*

1. *$u \in \mathcal{U}^{\mathbb{R}}$ is free; i.e. for all $u \in \mathcal{U}$, there exists a $y \in \mathcal{Y}^{\mathbb{R}}$ such that $(u, y) \in \mathcal{B}$.*
2. *$u \in \mathcal{U}^{\mathbb{R}}$ is maximally free; i.e. none of the components of $y \in \mathcal{Y}^{\mathbb{R}}$ can be chosen freely.*

The definition states that an input/output partition of the signal space is sensible only if the input u can be chosen freely (property 1) and if it determines univocally, once fixed the initial conditions, the output (property 2). In control theory a very essential and logical feature an I/O dynamical system must have is non anticipation. This property requires that the values of the output at a certain time does not depend on the future values of the input. An I/O dynamical system which enjoys this property is called *non anticipating I/O dynamical system.*

It is possible to combine the notions of input and output with the concept of state and to obtain a very useful representation of dynamical systems:

Definition 1.16 (I/S/O dynamical system). *An I/S/O dynamical system is defined as a 5-tuple $\Sigma_{I/S/O} = (\mathbb{R}, \mathcal{U}, \mathcal{Y}, \mathcal{X}, \mathcal{B}_f)$ where $\mathbb{R}$ represents the time axis, $\mathcal{U}$ the input signal space, $\mathcal{Y}$ the output signal space, $\mathcal{X}$ the state space and $\mathcal{B}_f \subset (\mathcal{U} \times \mathcal{Y}, \mathcal{X})^{\mathbb{R}}$ the full behavior. The following axioms must be satisfied:*

1. *$(\mathbb{R}, \mathcal{U} \times \mathcal{Y}, \mathcal{X}, \mathcal{B}_f)$ is a state space dynamical system*
2. *$(\mathbb{R}, \mathcal{U}, \mathcal{Y} \times \mathcal{X}, \mathcal{B}_f)$ is a non anticipating I/O dynamical system*
3. *$(\mathbb{R}, \mathcal{U}, \mathcal{X}, \mathcal{B}'_f)$ is a non anticipating I/O dynamical system. $\mathcal{B}'_f$ represents the input/state behavior:*
   $\mathcal{B}'_f = \{(u, x) : \mathbb{R} \to \mathcal{U} \times \mathcal{X} \mid \exists y : \mathbb{R} \to \mathcal{Y} \; s.t. \; (u, y, x) \in \mathcal{B}_f$

An I/S/O dynamical system is nothing else than a state space dynamical system (property 1) where an input output partition has been made on the signal space. Furthermore neither the output nor the state depend on future information of the input variable (properties 2 and 3).

Given an I/S/O dynamical system $\Sigma_{I/S/O} = (\mathbb{R}, \mathcal{U}, \mathcal{Y}, \mathcal{X}, \mathcal{B}_f)$ it is always possible to obtain an I/O dynamical system $\Sigma_{I/O} = (\mathbb{R}, \mathcal{U}, \mathcal{Y}, \mathcal{B})$ with:

$$\mathcal{B} = \{(u, y) \in (\mathcal{U} \times \mathcal{Y}) \mid \exists x \in \mathcal{X} \text{ s.t. } (u, y, x) \in \mathcal{B}_f\}$$

It is possible to prove that the behavioral equations that represent an I/S/O dynamical system are composed by *state evolution law* and by an output map and can be written in the form:

$$\begin{cases} f_1(\dot{x}) = f(x, u) \\ \\ h_1(y) = h(x, u) \end{cases} \tag{1.14}$$

which includes the well known form [128]:

$$\begin{cases} \dot{x} = f(x, u) \\ \\ y = h(x, u) \end{cases} \tag{1.15}$$

1.3 Physical Modeling

The framework introduced so far is very general and allows to model every phenomenon. However, in order to get a workable framework, it is necessary to add some structure to the models. Considering a certain class of phenomena, it is possible to obtain a class of models with a particular structure which can be exploited for various purposes: analysis of certain properties (e.g. observability, reachability), design of a suitable controller, etc.

Example 1.17. A very important class of dynamical systems is the one of linear systems. These systems can be modeled as I/S/O dynamical systems where both the state space and the input signal space and the output signal space are vector spaces and, furthermore, where the state evolution law and the output law are linear in the state and in the input. This leads to the following very well known equations:

$$\begin{cases} \dot{x} = Ax + Bu \\ \\ y = Cx + Du \end{cases} \tag{1.16}$$

where A, B, C and D are matrices of proper dimensions.

In the previous example, a mathematical property, the linearity, determines the kind of behavior of the system and gives a well defined structure to the behavioral equations. In this book, we are interested to modeling *physical systems* and, therefore, *physics* will determine the kind of behavior of the system. It is possible to describe physical system within the behavioral framework described in Sec. 1.2 and, furthermore, to exploit the features common to *all* physical system to determine their kind of behavior and to bring into evidence a lot of physical information into the behavioral equations.

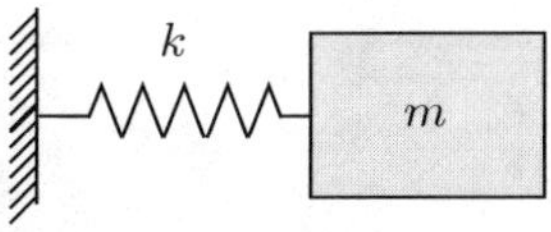

Fig. 1.3. A mass-spring system

The most important feature characterizing physical systems is the concept of *energy* and that it is possible to describe their dynamical behavior by means of *energy variables*, namely variables that are related to energy storage phenomena. A physical system has a dynamical behavior if and only if some energy exchange takes place either among the various parts that compose it or with the external world [237, 36, 141, 290] and, therefore, it is natural that energy must play a central role in the modeling process.

Consider the simple linear oscillator represented in Fig. 1.3 made up of a linear spring and of a mass characterized by stiffness k and inertia m respectively. Let x be the displacement of the spring and $p = mv_1$ be the momentum of the mass. It is well known from physics that the system is characterized by the following energy function:

$$E = \frac{1}{2}kx^2 + \frac{p^2}{2m} \tag{1.17}$$

Energy can be split in two contributes: $E_p = \frac{1}{2}kx^2$, which represents the potential energy stored by the spring, and $E_k = \frac{p^2}{2m}$, which represents the kinetic energy stored in the mass. Suppose that the system is oscillating, where does the motion derive from? The oscillator is the interconnection of two simpler subsystems: the mass and the spring. The situation is illustrated in Fig. 1.4.

The equations governing each subsystem are:

$$\begin{cases} \dot{p} = F_1 \\ v_1 = \frac{p}{m} \end{cases} \begin{cases} \dot{x} = v_2 \\ F_2 = kx \end{cases} \tag{1.18}$$

It is well known that $P = Fv$ represents mechanical power, namely the variation of mechanical energy, and thus, each subsystem can exchange energy with the rest of the world through a port defined by two variables: force and velocity. In fact, acting on force and velocity we can act on the power injected into (extracted from) the system and, therefore, on the amount of energy that is stored into (released by) the system. The interconnection, namely the exchange of information, takes place through force and velocity: the mass takes force as input and gives velocity as output while the spring takes a velocity as input and gives force as output. In other words, the interconnection determines the way in which energy is exchanged between the mass and the spring by imposing the following constraints on the port variables of each subsystem:

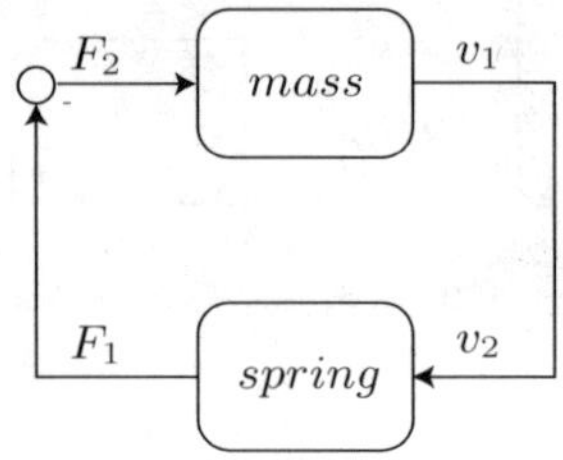

Fig. 1.4. Oscillator as the interconnection of mass and spring

$$\begin{cases} v_1 = v_2 \\ F_1 = -F_2 \end{cases} \tag{1.19}$$

Referring to Eq.(1.18) and to Eq.(1.19) it is possible to see that:

$$\begin{aligned} \frac{dE_k}{dt} &= \frac{\partial E_k}{\partial p}\dot{p} = \frac{p}{m}F_1 = -F_2 v_1 = -F_2 v_2 = -P_2 \\ \frac{dE_p}{dt} &= \frac{\partial E_p}{\partial x}\dot{x} = kxv_2 = F_2 v_2 = P_2 \end{aligned} \tag{1.20}$$

The amount of energy injected into the spring is exactly equal to the amount of energy extracted from the mass meaning that energy is simply transferred along the interconnection (or, in other words, that the interconnection is power preserving) and that, as well known from physics, the total energy E is constant.

Summarizing, each subsystem (mass and spring) is characterized by a variable (energy variable or energy state), that represents an energy storage phenomenon, and interacts with the rest of the world by exchanging energy through two variables (power variables) that represent force and velocity. By interconnecting the two subsystems, it is possible to obtain a new physical system (the oscillator) whose energy is the sum of the energy stored in the subsystems and whose dynamic behavior is determined by the mutual exchange of energy between the subsystems. The way in which energy is exchanged is given by the constraints imposed on the power variables of the interacting systems by the interconnection.

From the simple example of the oscillator, it is clear that in order to build an energy-based model for physical systems, it is useful to formally define the port through which a system can exchange energy with the rest of the world; this leads to the very important notion of *power port*.
To this aim, we need to introduce the concepts of dual space and of duality product:

Definition 1.18 (Dual Space). *Let $\mathcal{V}$ be a vector space, its dual space $\mathcal{V}^*$ is the set of linear maps from $\mathcal{V}$ to $\mathbb{R}$, i.e.:*

Table 1.1. Efforts and flows for various physical domains

Domain	Effort	Flow
Mechanics (translational)	Force F	Velocity v
Mechanics (rotational)	Torque τ	Angular Velocity ω
Electric	Voltage v	Current i
Hydraulic	Pressure p	Volume Flow Q
Thermodynamical	Temperature T	Entropy flow $\dot{E}$

$$\mathcal{V}^* = \{f : \mathcal{V} \to \mathbb{R} \mid \forall v_1, v_2 \in \mathcal{V} \alpha_1, \alpha_2 \in \mathbb{R} f(\alpha_1 v_1 + \alpha_2 v_2) = \alpha_1 f(v_1) + \alpha_2 f(v_2)\} \tag{1.21}$$

It is possible to show that $\mathcal{V}^*$ is a vector space and that if $\mathcal{V}$ is finite dimensional, then $\mathcal{V}^*$ has the same dimension of $\mathcal{V}$ [117]. Elements of $\mathcal{V}$ and $\mathcal{V}^*$ are said to be dual one with respect to the other. Furthermore elements of $\mathcal{V}$ are called *vectors* while elements of $\mathcal{V}^*$ are called *covectors*.

Definition 1.19 (Duality Product). *Given $v \in \mathcal{V}$ and $v^* \in \mathcal{V}^*$, the duality product is defined as:*

$$\langle , \rangle : \mathcal{V} \times \mathcal{V}^* \to \mathbb{R} \quad \langle v, v^* \rangle = v^*(v) \tag{1.22}$$

The duality product is *intrinsically* defined for any vector space and it is NOT an extra structure that can be associated to it. Once coordinates have been fixed, vectors and covectors are represented as column vectors and row vectors respectively. The duality product is simply given by the usual product between row and column vectors.

In each physical domain there is a pair of dual variables, called *power conjugated variables*, whose duality product represents power. These variables are generically called *flows* and *efforts* and live in dual vector spaces, the space of flows $\mathcal{F}$ and the space of efforts $\mathcal{E} = \mathcal{F}^*$. In Tab. 1.1 effort and flow variables for various physical domains are reported.
We are now ready to introduce power ports.

Definition 1.20 (Power port). *Let $\mathcal{F}$ and $\mathcal{E} = \mathcal{F}^*$ be the flow and effort vector spaces. A power port is defined as $P = \mathcal{F} \times \mathcal{E}$. Given $f \in \mathcal{F}$ and $e \in \mathcal{E}$, the product $\langle e, f \rangle$ is called the power (traversing the power port), where $\langle , \rangle$ is the intrinsic duality product defined in Def. 1.19.*

Power ports are the medium through which a physical system can exchange energy with the rest of the world and, in particular, through which it can be interconnected with other physical systems. In the simple example of the oscillator, the power ports of the mass and of the spring are the pairs (F_1, v_1) and (F_2, v_2) respectively and they are used to interconnect the two systems.

The fact that power conjugated variables are independent of the configuration of the physical system and that the duality product is defined intrinsically

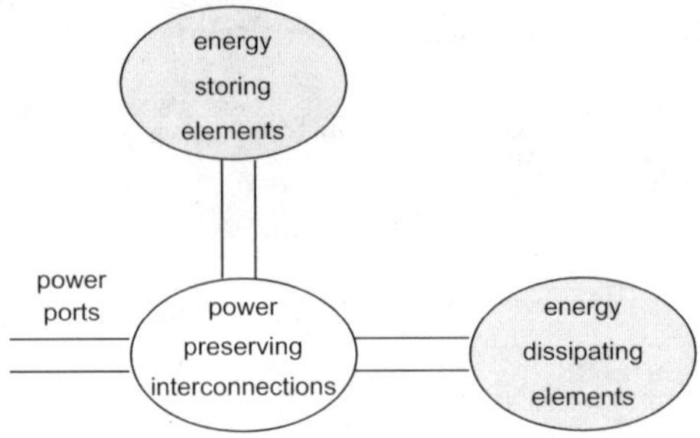

Fig. 1.5. Network model of a physical system

is of crucial importance. This allows to uniquely define the energy exchange without any additional structure on the port and to use power ports to interconnect physical systems characterized by different configuration spaces and even by different physical domains. Roughly speaking, a power port is an interface through which energy flows and this flow is expressed in an intrinsic way, independently of the physical domain and of the particular configuration of the system the power port is associated to.

1.4 Implicit Port-Hamiltonian Systems

From the very simple example proposed in the previous section, it is evident that three concepts are fundamental for building the model of the oscillator reported in Fig. 1.3: the state space, whose components model the energy storage phenomena associated to the constitutive parts of the system, the power port, the medium through which each physical subsystem can exchange energy, and the interconnection structure, namely the way by which all the subsystems are joined through their power ports and, consequently, by which they exchange energy.

The behavior of any physical system can be described in terms of energy storage and energy flows. The aim of this section is to describe the port-Hamiltonian formalism, introduced in [189, 323, 322, 318, 319], which allows to describe a physical system in terms of energy exchange and of interconnection of basic elements and to give it a behavioral interpretation.

Following the network paradigm [237, 36, 141], a physical system can be described by a set of elements *storing kinetic energy*, a set of elements *storing potential energy*, a set of elements *dissipating free energy* and a set of *power ports* (by means of which interaction with the external environment or interconnection with other systems can take place) interconnected with each other by means of *power preserving interconnections*, as illustrated in Fig. 1.5. The fact that the various elements are joined with a power preserving interconnection means that along the interconnections there can *only* be energy transfer and *NOT* energy production. This allows the presence of power pre-

serving elements such as transformers and gyrators in the interconnection. Associated to the energy storing elements there are energy variables $x_1, \ldots, x_n$, being coordinates for some n-dimensional smooth manifold $\mathcal{X}$, and a function $H : \mathcal{X} \to \mathbb{R}$ representing the total stored energy. Keeping in mind the network modeling point of view, it is possible to introduce what is called an *implicit port-Hamiltonian system*, which is the mathematical formalization of a network model of a physical system.
Four ingredients are needed to define an [illegible]it port-Hamiltonian system:

1. A state space $\mathcal{X}$
2. A space of flow variables, assumed to be a vector space, denoted by $\mathcal{V}$. The dual space of effort variables, denoted by $\mathcal{V}^*$.
3. A geometric structure $\mathcal{D}$, called Dirac structure, which represents the "energetic topology" of the system, i.e. how the energy flows among the various parts of the system.
4. A smooth function $\mathcal{H}$, called Hamiltonian function, defined on the state space and expressing the total energy of the system corresponding to a certain state.

The state space $\mathcal{X}$ is a smooth manifold whose elements are energy variables which represent the amount of total energy stored into the system. The dual spaces of flow and effort variables are used to define, with their intrinsically defined duality product, power ports by means of which it is possible to interact with the system or to interconnect it with other systems. These vector spaces are independent of the configuration in order to allow interconnection between systems with different configuration spaces. The Dirac structure is the mathematical object used to describe the internal power preserving interconnections between the various elements composing the system. Finally, the Hamiltonian function expresses the energy of the system given a certain configuration in the state manifold.

Consider the state manifold $\mathcal{X}$ and its tangent and cotangent bundles, denoted by $T\mathcal{X}$ and $T^*\mathcal{X}$ respectively. Given a point $x \in \mathcal{X}$, the tangent space in x, $T_x\mathcal{X}$ (whose elements are vectors), and the cotangent space, $T_x^*\mathcal{X}$ (whose elements are covectors), are dual vector spaces by definition and, therefore, there exists an intrinsic duality product on $\mathcal{X}$, denoted by $\langle , \rangle$, being a bilinear map from $T\mathcal{X} \times T^*\mathcal{X}$ to $\mathbb{R}$, defined by

$$\langle v^*, v \rangle = v^*(v) \qquad v \in T_x\mathcal{X}, v^* \in T_x^*\mathcal{X}, x \in \mathcal{X} \tag{1.23}$$

Summarizing, the vector spaces $\mathcal{V}$ and $\mathcal{V}^*$ are used to define the power ports, the way by which the port-Hamiltonian system can interact, by exchanging energy, with other systems. The tangent and cotangent bundles $T\mathcal{X}$ and $T^*\mathcal{X}$ of the state manifold $\mathcal{X}$ are used to define the way by which internal energy exchange takes place (a kind of *internal*, and, therefore, state dependent, power ports). The Dirac structure describes *how* this energy exchange takes place.

Consider the following vector bundle over the manifold $\mathcal{X}$ (see Sec. A.1):

$$\mathcal{Q} = T\mathcal{X} \oplus T^*\mathcal{X} \oplus \mathcal{V} \oplus \mathcal{V}^* \tag{1.24}$$

The fibers of the bundle $\mathcal{Q}$ define the space of vectors, covectors, flows and efforts:

$$\mathcal{Q}(x) = T_x\mathcal{X} \times T_x^*\mathcal{X} \times \mathcal{V} \times \mathcal{V}^* \tag{1.25}$$

There exists a symmetric bilinear non degenerate form on $\mathcal{Q}$, called the + pairing, defined by:

$$\begin{aligned} &\langle , \rangle_+ : \mathcal{Q}(x) \times \mathcal{Q}(x) \to \mathbb{R} \\ &\langle (v, v^*, f, e), (w, w^*, \tilde{f}, \tilde{e}) \rangle_+ = \langle v^*, w \rangle + \langle w^*, v \rangle + \langle e, \tilde{f} \rangle + \langle \tilde{e}, f \rangle \end{aligned} \tag{1.26}$$

where $(v, v^*, f, e), (w, w^*, \tilde{f}, \tilde{e}) \in \mathcal{Q}(x)$.

Definition 1.21 (Dirac structure). *A Dirac structure on the bundle $\mathcal{Q}$ is a smooth vector sub-bundle $\mathcal{D} \subset \mathcal{Q}$ such that for every $x \in \mathcal{X}$, the fiber $\mathcal{D}(x) \subset \mathcal{Q}(x)$ satisfies the condition $\mathcal{D}(x) = \mathcal{D}^\perp(x)$, where:*

$$\begin{aligned} &\mathcal{D}^\perp(x) = \\ &= \{(w, w^*, \tilde{f}, \tilde{e}) \in \mathcal{Q}(x) \mid \langle (v, v^*, f, e), (w, w^*, \tilde{f}, \tilde{e}) \rangle_+ = 0, \forall (v, v^*, f, e) \in \mathcal{D}(x)\} \end{aligned} \tag{1.27}$$

It follows directly from the definition that:

$$\langle v, v^* \rangle + \langle e, f \rangle = \frac{1}{2} \langle (v, v^*, f, e), (v, v^*, f, e) \rangle_+ = 0 \quad \forall (v, v^*, f, e) \in \mathcal{D}(x) \tag{1.28}$$

Thus a Dirac structure represents a power preserving relation between the internal and external power variables and therefore it can be profitably used to describe the topology, i.e. the internal interconnection structure, of any physical system. Loosely speaking, the set of power preserving interconnections along which the elements constituting the physical system exchange energy is represented by the Dirac structure.

It is now possible to introduce implicit port-Hamiltonian systems. Assume, for the moment, that there is no dissipation in the system.

Definition 1.22 (Implicit port-Hamiltonian system). *Consider a state space $\mathcal{X}$, a space of flow variables $\mathcal{V}$ and, dually, the space of effort variables $\mathcal{V}^*$, a Dirac structure $\mathcal{D}$ and a Hamiltonian function $\mathcal{H}$. Then, the implicit port-Hamiltonian system corresponding to the 4-tuple $(\mathcal{X}, \mathcal{V}, \mathcal{D}, \mathcal{H})$ is defined by setting:*

$$v = -\dot{x} \quad \textit{and} \quad v^* = \frac{\partial \mathcal{H}}{\partial x}(x) \tag{1.29}$$

That is it is defined by the equations

$$(-\dot{x}, \frac{\partial \mathcal{H}}{\partial x}, f, e) \in \mathcal{D}(x) \tag{1.30}$$

The property reported in Eq.(1.28) immediately yields the following power balance:

$$\dot{\mathcal{H}}(x(t)) = \langle \frac{\partial \mathcal{H}}{\partial x}(x(t)), \dot{x}(t) \rangle = \langle e(t), f(t) \rangle \tag{1.31}$$

that is implicit port-Hamiltonian systems are *lossless*, i.e. all the energy that is supplied through the power ports ($\langle e(t), f(t) \rangle$) is stored into the system. It is possible to include dissipation in the implicit port-Hamiltonian framework by simply terminating some of the power ports with dissipating elements, namely elements that always extract energy from those ports. Such a kind of element can be described by the following constitutive relation

$$e_r = \alpha(f_r) \quad \textit{where} \quad \alpha : \mathcal{F} \rightarrow \mathcal{E} \text{ such that} \quad \langle e_r, f_r \rangle \leq 0 \tag{1.32}$$

These elements supply always a negative power to the system through the power port they are connected to and, therefore, they always extract energy from the system.

An implicit port-Hamiltonian system where some of the power ports are terminated with energy absorbing elements is called an *implicit port-Hamiltonian system with dissipation* and for these systems the power balance of Eq.(1.31) becomes:

$$\dot{\mathcal{H}}(x(t)) = \langle e(t), f(t) \rangle + \langle e_r(t), f_r(t) \rangle \leq \langle e(t), f(t) \rangle \tag{1.33}$$

that is, an implicit port-Hamiltonian system with dissipation is *passive*, namely the energy supplied through the power ports is either stored or dissipated.

A very clear interpretation of implicit port-Hamiltonian systems can be given in terms of the behavioral approach. Energy plays a central role in modeling physical systems and energy exchange determines their evolution in time. Because of their evolution in time, physical systems can be modeled as a state space dynamical systems, defined in Def. 1.13. Considering lumped parameters continuous systems, $\mathbb{R}$ can be chosen as time axis. It is very natural to choose as manifest variables those characterizing the external interface of the system with the rest of the world, that is those representing the power ports. Thus, the manifest signal space can be chosen as $\mathcal{W} = \mathcal{V} \times \mathcal{V}^*$, where $\mathcal{V}$ and $\mathcal{V}^*$ represent the space of flows and the space of efforts respectively. The memory of a physical system is represented by the evolution of the storage of energy, and, therefore, it is described by the energy variables which live on a smooth manifold $\mathcal{X}$ and by the energy function $\mathcal{H}$. The last, and most crucial point, is the definition of the behavior. To define the behavior means to define a relation between the state evolution, the state itself and the manifest variables. For physical systems, this relation is given by their network structure: the behavior must represent an energy exchange through power preserving interconnection. It is necessary then to define a function representing the energy of the system and a Dirac structure representing the internal power preserving interconnections. The relation that expresses the behavior of a physical system is the one reported in Eq.(1.30).

As linearity on the manifest variables signal space and on the state space was the structure to impose to obtain the class of linear systems, as shown in Example 1.17, the energy function, the Dirac structure and the power port are the entities to consider to model the class of physical systems. Nevertheless, energy, despite of linearity, is a property shared by all physical systems and, therefore, the port-Hamiltonian framework allows to describe physical systems independently of their complexity.

In implicit port-Hamiltonian models the physics of the system, that is energy flows and interconnection structure, is not hidden but it is explicitly shown. Furthermore this formalism allows to describe both differential relations, arising because of the system evolution, and algebraic relations, arising because of the network structure of any physical system. Finally, no causality has been fixed. This is very important since, as already pointed out, very often the input/output selection depends on the particular task a system is executing.

1.4.1 A Coordinate Based Representation

The definition of Dirac structure and of implicit port-Hamiltonian systems have been given in a coordinate free way. This has been possible since these concepts relies on intrinsic and coordinate independent properties that are common to all physical systems. On the other hand, it is often very useful to represent the dynamic behavior of a system by means of differential equations.

There are several ways of representing a Dirac structure and consequently of writing equations to model an implicit port-Hamiltonian system. Each representation has its own advantages and describes in a particular way the power preserving interconnection of a physical system. In this section it will be illustrated the so-called *Kernel representation* [319] which will be used through the whole book. For other representations, see [66, 25, 319].

Consider an n-dimensional smooth manifold $\mathcal{X}$ and an m-dimensional vector space $\mathcal{V}$. Each Dirac structure $\mathcal{D}$ defined on the bundle $\mathcal{Q} = T\mathcal{X} \oplus T^*\mathcal{X} \oplus \mathcal{V} \oplus \mathcal{V}^*$ as in Def. 1.21 defines the smooth distributions:

$$
\begin{aligned}
G_0 &= \{(v, f) \in T\mathcal{X} \oplus \mathcal{V} \mid (v, 0, f, 0) \in \mathcal{D}\} \\
G_1 &= \{(v, f) \in T\mathcal{X} \oplus \mathcal{V} \mid \exists (v^*, e) \in T^*\mathcal{X} \oplus \mathcal{V}^* \text{ such that } (v, v^*, f, e) \in \mathcal{D}\}
\end{aligned}
\tag{1.34}
$$

and the smooth codistributions:

$$P_0 = \{(v^*, e) \in T^*\mathcal{X} \oplus \mathcal{V}^* \mid (0, v^*, 0, e) \in \mathcal{D}\}$$

$$P_1 = \{(v^*, e) \in T^*\mathcal{X} \oplus \mathcal{V}^* \mid \exists (v, f) \in T\mathcal{X} \oplus \mathcal{V} \text{ such that } (v, v^*, f, e) \in \mathcal{D}\} \tag{1.35}$$

It is possible to define for any smooth distribution G the smooth codistribution ann G as:

$$ann\, G = \{(v^*, e) \in T^*\mathcal{X} \oplus \mathcal{V}^* \mid \langle v^*, v\rangle + \langle e, f\rangle = 0\ \forall (v, f) \in G\} \tag{1.36}$$

and for any smooth codistribution P the smooth distribution ker P as:

$$ker\, P = \{(v, f) \in T\mathcal{X} \oplus \mathcal{V} \mid \langle v^*, v\rangle + \langle e, f\rangle = 0\ \forall (v^*, f) \in P\} \tag{1.37}$$

The following proposition holds [322]:

Proposition 1.23. *Let $\mathcal{D}$ be a Dirac structure on $\mathcal{Q}$ and G_0, G_1, P_0 and P_1 defined as in Eq.(1.34) and Eq.(1.35). Then:*

1. $G_0 = ker\, P_1$, $P_0 = ann\, G_1$
2. $P_1 \subset ann\, G_0$, $G_1 \subset ker\, P_0$, *with equality if G_1, respectively P_1 is constant dimensional.*

Proof.

1. $(v, f) \in G_0$ if and only if $(v, 0, f, 0) \in \mathcal{D}$ which means that:

$$\langle 0, w\rangle + \langle w^*, v\rangle + \langle 0, \tilde{f}\rangle + \langle \tilde{e}, f\rangle = 0 \quad \forall (w, w^*, \tilde{f}, \tilde{e}) \in \mathcal{D} \tag{1.38}$$

or, equivalently

$$\langle w^*, v\rangle + \langle \tilde{e}, f\rangle = 0 \quad \forall (w^*, \tilde{e}) \in P_1 \tag{1.39}$$

Thus $G_0 = ker\, P_1$. Similarly, $(v^*, e) \in P_0$ if and only if $(0, v^*, 0, e) \in \mathcal{D}$ which means that:

$$\langle v^*, w\rangle + \langle w^*, 0\rangle + \langle e, \tilde{f}\rangle + \langle \tilde{e}, 0\rangle = 0 \quad \forall (w, w^*, \tilde{f}, \tilde{e}) \in \mathcal{D} \tag{1.40}$$

or, equivalently

$$\langle v^*, w\rangle + \langle e, \tilde{f}\rangle = 0 \quad \forall (w, \tilde{f}) \in G_1 \tag{1.41}$$

Thus $P_0 = ann\, G_1$.

2. This follows from property 1 and from the inequalities $P \subset ann\, ker\, P$, $G \subset ker\, ann\, G$, that hold for any smooth distribution and codistribution P and G, with equality if P and G are constant-dimensional [219]

The distribution G_1 and the codistribution P_1 have a very clear interpretation when considering the implicit port-Hamiltonian system represented in Eq.(1.30) and corresponding to the Dirac structure $\mathcal{D}$ and to an energy function H. In this case the distribution G_1 describe the set of admissible flows $(\dot{x}, f)$ and, dually, the codistribution P_1 describe the set of algebraic constraints of the system, i.e.

$$(\frac{\partial H}{\partial x}, e) \in P_1$$

Definition 1.24 (Regular point for a Dirac structure). *A point $x \in \mathcal{X}$ is a regular point for a Dirac structure $\mathcal{D}$ on $\mathcal{Q}$ if the dimension of G_1 and P_1 (and, consequently, thanks to Proposition 1.23, of G_0 and P_0) is constant in a neighborhood of x*

The following proposition can be proven [322]:

Proposition 1.25. *For each regular point $x \in \mathcal{X}$,* $\dim \mathcal{D}(x) = \dim \mathcal{X} + \dim \mathcal{V}$

Proof. Since $\mathcal{X}$ is a smooth manifold $\forall x \in \mathcal{X}$, $\dim T_x\mathcal{X} = \dim T_x^*\mathcal{X} = \dim \mathcal{X}$ and, furthermore, $\dim \mathcal{V} = \dim \mathcal{V}^*$. It follows that for each fiber $K(x)$ of the bundle $\mathcal{Q}$:

$$\begin{aligned} &\dim K(x) = \dim(T_x\mathcal{X} \times T_x^*\mathcal{X} \times \mathcal{V} \times \mathcal{V}^*) = \dim T_x\mathcal{X} + \dim T_x^*\mathcal{X} + \\ &+ \dim \mathcal{V} + \dim \mathcal{V}^* = 2\dim \mathcal{X} + 2\dim \mathcal{V} \end{aligned} \tag{1.42}$$

Since the +-pairing is a non degenerate bilinear form at any regular point, if $S \subset T_x\mathcal{X} \times T_x^*\mathcal{X} \times \mathcal{V} \times \mathcal{V}^*$, then $\dim S + \dim S^\perp = 2\dim \mathcal{X} + 2\dim \mathcal{V}$. Since, by definition, $\mathcal{D}(x) = \mathcal{D}^\perp(x)$ it is possible to write:

$$\dim \mathcal{D}(x) + \dim \mathcal{D}^\perp(x) = 2\dim \mathcal{D} = 2\dim \mathcal{X} + 2\dim \mathcal{V} \tag{1.43}$$

and thus:

$$\dim \mathcal{D}(x) = \dim \mathcal{X} + \dim \mathcal{V} \tag{1.44}$$

Since the set of regular points is open and dense in $\mathcal{X}$ and $\mathcal{D}$ is a vector sub-bundle, it follows that:

$$\dim \mathcal{D}(x) = \dim \mathcal{X} + \dim \mathcal{V} = n + m \quad \forall x \in \mathcal{X} \tag{1.45}$$

Furthermore, since $\mathcal{D}$ is a smooth vector sub-bundle, it is possible to find locally about every point $x \in \mathcal{X}$ two $(n+m) \times (n+m)$ matrices $F(x)$ and $E(x)$ that depends smoothly on x and such that, locally:

$$\begin{aligned} &\mathcal{D}(x) = \{(v, v^*, f, e) \in T_x\mathcal{X} \times T_x^*\mathcal{X} \times \mathcal{V} \times \mathcal{V}^* \mid F(x)\begin{pmatrix} v \\ f \end{pmatrix} + E(x)\begin{pmatrix} v^* \\ e \end{pmatrix} = 0\} \\ &rank[F(x) : E(x)] = n + m \end{aligned} \tag{1.46}$$

Moreover, since $\mathcal{D} = \mathcal{D}^\perp$, the following relation necessarily holds:

$$E(x)F^T(x) + F(x)E^T(x) = 0 \tag{1.47}$$

Matrices $E(x)$ and $F(x)$ such that Eq.(1.46) and Eq.(1.47) holds are commonly called a *Kernel representation* of a Dirac structure $\mathcal{D}$. Considering the definition of an implicit port-Hamiltonian system given in Eq.(1.30), it is possible to find a kernel representation of an implicit port-Hamiltonian system, that can be written as:

$$F(x)\begin{pmatrix}-\dot{x}\\ f\end{pmatrix}+E(x)\begin{pmatrix}\frac{\partial H}{\partial x}\\ e\end{pmatrix}=0 \tag{1.48}$$

where the matrices $E(x)$ and $F(x)$ satisfies Eq.(1.46) and Eq.(1.47) for the Dirac structure corresponding to the implicit port-Hamiltonian system.

Remark 1.26. The crucial requirement for the formulation of the kernel representation is the existence of a regular point on the state manifold $\mathcal{X}$. This requirement is very often satisfied in practice.

In order to include dissipation, some ports have to be terminated by dissipative elements. Thus it is possible to split the power conjugated variable in two parts:

$$f=(f_P,f_R) \qquad e=(e_P,e_R) \tag{1.49}$$

where the subscript P denotes the part of power variables associated to power ports for the interaction with the external world, while the subscript R denotes the part of power variables associated to a dissipative elements. It is possible to partition matrices $E(x)$ and $F(x)$ in order to make explicit the presence of dissipation and to get a kernel representation of an implicit port-Hamiltonian system with dissipation:

$$\underbrace{\left(F_S(x)\ F_P(x)\ F_R(x)\right)}_{F(x)}\begin{pmatrix}-\dot{x}\\ f_P\\ f_R\end{pmatrix}+\underbrace{\left(E_S(x)\ E_P(x)\ E_R(x)\right)}_{E(x)}\begin{pmatrix}\frac{\partial H}{\partial x}\\ e_P\\ e_R\end{pmatrix}=0 \tag{1.50}$$

where the matrices $F(x)$ and $E(x)$ have been partitioned in a part relative to the energy storage, a part relative to the power ports for the interaction with the external world and a part relative to energy dissipation.

In case there are no algebraic constraints on the state variables, it is possible to see [319] that the matrices $F(x)$ and $E(x)$ assume the following special form:

$$\begin{aligned}
F_S(x)&=\begin{pmatrix}I\\0\\0\end{pmatrix} \quad F_P(x)=\begin{pmatrix}g(x)\\0\\0\end{pmatrix} \quad F_R(x)=\begin{pmatrix}g_R(x)\\0\\0\end{pmatrix}\\
E_S(x)&=\begin{pmatrix}J(x)\\-g_R^T(x)\\-g^T(x)\end{pmatrix} \quad E_P(x)=\begin{pmatrix}0\\0\\I\end{pmatrix} \quad E_R=\begin{pmatrix}0\\I\\0\end{pmatrix}
\end{aligned} \tag{1.51}$$

where $J(x)$ is a skew-symmetric matrix and I represents the identity matrix of proper dimension.

Very often it is possible to model dissipation with elements characterized by the relation

$$f_R = -\tilde{R}(x)e_R \tag{1.52}$$

where $\tilde{R}(x)$ is a symmetric positive semidefinite matrix. Thus, substituting Eq.(1.52) in Eq.(1.51), the following equations can be obtained:

$$-\dot{x} + g(x)f_P + g_R(x)f_R + J(x)\frac{\partial H}{\partial x} = 0 \tag{1.53}$$

$$-g_R^T(x)\frac{\partial H}{\partial x} + e_R = 0 \tag{1.54}$$

$$-g^T(x)\frac{\partial H}{\partial x} + e_p = 0 \tag{1.55}$$

By substituting Eq.(1.54) and Eq.(1.52) in Eq.(1.53), we obtain the following equation:

$$\dot{x} = J(x)\frac{\partial H}{\partial x} - \underbrace{g_R^T(x)\tilde{R}(x)g_R(x)}_{R(x)}\frac{\partial H}{\partial x} + g(x)f_P \tag{1.56}$$

where $R(x) = g_R^T(x)\tilde{R(x)}g_R(x)$ is a symmetric positive semidefinite matrix representing the dissipation of the system. It is now possible to give the following definition:

Definition 1.27 (Explicit port-Hamiltonian systems). *An explicit port-Hamiltonian system is an I/S/O continuous time dynamical system defined by the 5-tuple* $(\mathbb{R}, \mathcal{U} \times \mathcal{Y}, \mathcal{X}, \mathcal{B}_f)$ *where:*

- $\mathcal{X}$ *is an n dimensional manifold representing the state space; the states are energy variables*
- $\mathcal{U}$ *is the input vector space; the input is a power conjugated variable.*
- $\mathcal{Y} = \mathcal{U}^*$ *is the output vector space; the output is a power variable dual to the input.*
- $\mathcal{B}_f$ *is the full behavior of the system.*

The full behavior is represented by the following equations:

$$\begin{cases} \dot{x} = [J(x) - R(x)]\frac{\partial H}{\partial x} + g(x)u \\ y = g^T(x)\frac{\partial H}{\partial x} \end{cases} \tag{1.57}$$

where $J(x)$ *is a skew symmetric matrix representing the internal power preserving interconnections,* $R(x)$ *is a symmetric positive semidefinite matrix representing the dissipation of the system and* $g(x)$ *is a matrix describing the way that power coming from the external world is distributed into the system.*

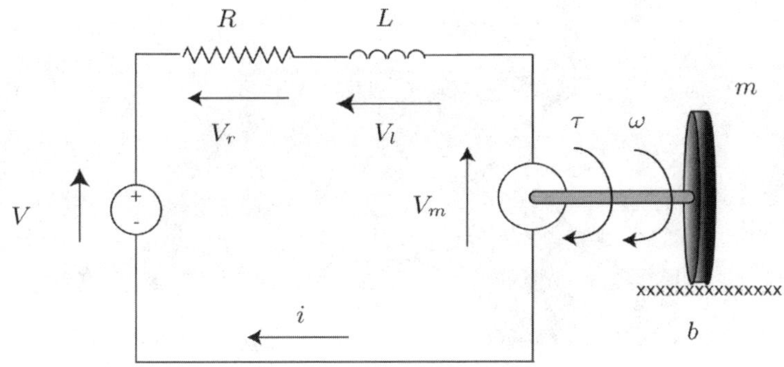

Fig. 1.6. Schematic representation of a DC motor

Notice that the representation in Eq.(1.57) satisfies Theorem 1.14 and thus, it properly defines an I/S/O dynamical system.

Remark 1.28. Through the book with port-Hamiltonian system is meant *explicit* port-Hamiltonian system. When referring to non explicit port-Hamiltonian systems the word *implicit* will be always used.

Example 1.29 (DCMotor). Consider the schematic representation of a DC motor given in Fig. 1.6. The energy variables are the flux ϕ on the inductor and the momentum p of the load; thus the state space is $\mathbb{R}^2$. The energy function is given by:

$$H : \mathbb{R}^2 \to \mathbb{R} \qquad H = \frac{\phi^2}{2L} + \frac{p^2}{2m}$$

where L and m represents the inductance of the circuit and the inertia of the load respectively. The dissipative elements can be modeled by the following constitutive law:

$$f_R = -\tilde{R} e_R \qquad \tilde{R} = \begin{pmatrix} R & 0 \\ 0 & b \end{pmatrix}$$

where R and b represent the resistance of the circuit and the damping on the load respectively. The port (manifest) variables are the input voltage (which is an effort) and the relative port current (which is a flow).
Defining:

$$x = \begin{pmatrix} \phi \\ p \end{pmatrix}$$

and letting k represent the electro-mechanical coupling, a kernel representation of the port-Hamiltonian model of the system is the following:

$$\begin{pmatrix} 1 & 0 & 0 & 1 & 0 \\ 0 & 1 & 0 & 0 & 1 \\ 0 & 0 & 0 & 0 & 0 \\ 0 & 0 & 1 & 0 & 0 \\ 0 & 0 & 0 & 1 & 0 \end{pmatrix} \begin{pmatrix} -\dot{x} \\ f_P \\ f_R \end{pmatrix} + \begin{pmatrix} 0 & -k & 1 & 0 & 0 \\ k & 0 & 0 & 0 & 0 \\ -1 & 0 & 0 & 1 & 0 \\ 0 & -1 & 0 & 0 & 1 \\ -1 & 0 & 0 & 0 & 0 \end{pmatrix} \begin{pmatrix} -\frac{\partial H}{\partial x} \\ e_P \\ e_R \end{pmatrix} \tag{1.58}$$

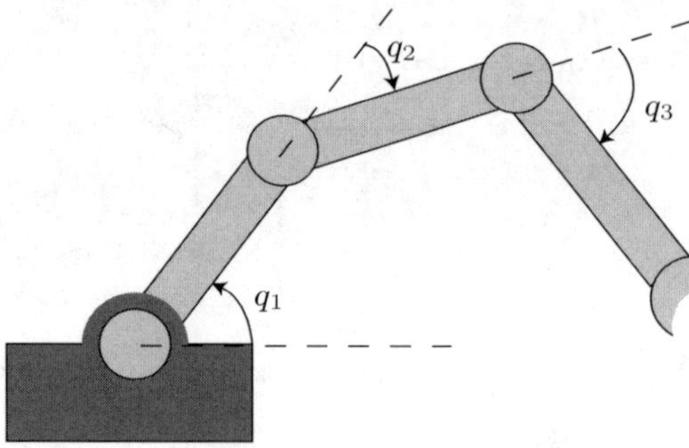

Fig. 1.7. Schematic representation of a n-DOF anthropomorphic robot

The model of the DC motor does not contain constraints on the state variables and, thus, it is possible to give an explicit port-Hamiltonian model of the system:

$$\begin{cases} \begin{pmatrix} \dot{\phi} \\ \dot{p} \end{pmatrix} = \left[\begin{pmatrix} 0 & -k \\ k & 0 \end{pmatrix} - \begin{pmatrix} R & 0 \\ 0 & b \end{pmatrix} \right] \begin{pmatrix} \frac{\partial H}{\partial \phi} \\ \frac{\partial H}{\partial p} \end{pmatrix} + \begin{pmatrix} 1 \\ 0 \end{pmatrix} e_P \\ f = (1\ 0) \begin{pmatrix} \frac{\partial H}{\partial \phi} \\ \frac{\partial H}{\partial p} \end{pmatrix} \end{cases} \tag{1.59}$$

where:

$$J(x) = \begin{pmatrix} 0 & -k \\ k & 0 \end{pmatrix} \qquad R(x) = \begin{pmatrix} R & 0 \\ 0 & b \end{pmatrix}$$

The matrix $J(x)$ is constant and represents the energetic interconnection due to the coupling between the electrical and the mechanical part. The matrix $R(x)$ is positive definite and represents the dissipation present on the system. The port effort e represents the input voltage and the output flow f is the related current.

Example 1.30 (n-DOF robot). Consider an n-DOF fully-actuated mechanical system with generalized coordinates $q \in \mathcal{Q}$; in Fig. 1.7 an example of a 3-DOF robot is reported. If $p = M(q)\dot{q} \in T_q^*\mathcal{Q}$ are the generalized momenta, with $M(q)$ the inertia matrix, an explicit port-Hamiltonian representation of this system can be obtained by assuming in $\dim \mathcal{X} = 2n$ and $m = n$, then defining $x := (q\ p)^T$, $H(q,p) := \frac{1}{2}p^T M^{-1}(q)p + V(q)$, where $V(q)$ is the potential energy, and, finally,

$$J = \begin{pmatrix} 0 & I_n \\ -I_n & 0 \end{pmatrix} \; R = \begin{pmatrix} 0 & 0 \\ 0 & D(q,p) \end{pmatrix} \; G = \begin{pmatrix} 0 \\ B(q) \end{pmatrix}$$

with $D(q,p) = D^T(q,p) \geq 0$ taking into account the dissipation effects. Moreover, assume $\operatorname{rank} G = n$, since the mechanical system is fully actuated. Finally the input is an effort representing the input torques and the output is a flow representing the joint velocity. These considerations lead to the following model

$$\begin{cases} \begin{pmatrix} \dot{q} \\ \dot{p} \end{pmatrix} = \left[\begin{pmatrix} 0 & I_n \\ -I_n & 0 \end{pmatrix} - \begin{pmatrix} 0 & 0 \\ 0 & D(q,p) \end{pmatrix} \right] \begin{pmatrix} \frac{\partial H}{\partial q} \\ \frac{\partial H}{\partial p} \end{pmatrix} + \begin{pmatrix} 0 \\ B \end{pmatrix} e \\ f = \begin{pmatrix} 0 & B^T \end{pmatrix} \begin{pmatrix} \frac{\partial H}{\partial q} \\ \frac{\partial H}{\partial p} \end{pmatrix} \end{cases} \tag{1.60}$$

1.5 Geometric Scattering

In Sec. 1.3 it has been shown that in any physical domain there exists two power conjugated variables whose intrinsically defined dual product gives power. This observation led to the definition of power port, the medium through which each physical system can exchange energy with the rest of the world.

The exchange of energy through the power port can be split in two contributions: an incoming power flow and an outgoing power flow that represent the power that is supplied to the system and the power that is extracted from the system respectively ; this splitting is not intrinsic but it depends on some parameters representing an impedance. This observation leads to a very useful and interesting representation of power port: the *scattering representation.*

Consider a power port $P = (\mathcal{V} \times \mathcal{V}^*)$ where the $\mathcal{V}$ and $\mathcal{V}^*$ are dual vector spaces representing the space of efforts and the space of flow respectively. Suppose that $\dim \mathcal{V} = \dim \mathcal{V}^* = n$. It is possible to define, analogously to what has been done in Eq.(1.26) on a vector bundle, a + pairing, namely a symmetric bilinear non degenerate form, on $\mathcal{V} \times \mathcal{V}^*$. This operator is defined by the following relations:

$$\begin{aligned} &\langle , \rangle_+ : (\mathcal{V} \times \mathcal{V}^*) \times (\mathcal{V} \times \mathcal{V}^*) \to \mathbb{R} \\ &\langle (f_1, e_1), (f_2, e_2) \rangle_+ = \langle e_2, f_1 \rangle + \langle e_1, f_2 \rangle \qquad (f_1, e_1), (f_2, e_2) \in \mathcal{V} \times \mathcal{V}^* \end{aligned} \tag{1.61}$$

Consider a basis $\{e_1, \ldots, e_n\}$ of $\mathcal{V}$ and consider its corresponding representing matrix:

$$B = (e_1 \ldots e_n) \tag{1.62}$$

The dual basis, a basis of $\mathcal{V}^*$, $\{e_1^*, \ldots, e_n^*\}$ is represented by the matrix:

$$B_* = (e_1^* \ldots e_n^*) \tag{1.63}$$

such that $B_*^T B = B^T B_* = I$. It is thus possible to define a basis matrix for the space $\mathcal{V} \times \mathcal{V}^*$ by the following:

$$\bar{B} = \begin{pmatrix} B & 0 \\ 0 & B_* \end{pmatrix} \tag{1.64}$$

Finally, it is possible to define the dual matrix of $\bar{B}$:

$$\bar{B}_* = \begin{pmatrix} B_* & 0 \\ 0 & B \end{pmatrix} \tag{1.65}$$

Using the basis matrices introduced so far, a representation of the + pairing defined in Eq.(1.61) is given by:

$$T_{ij} = \bar{B}_* \begin{pmatrix} 0 & I \\ I & 0 \end{pmatrix} \bar{B}_*^T \tag{1.66}$$

The two indexes indicate that T is a $(0,2)$ tensor (see Sec. A.2). In order to define geometric scattering in a coordinate free way it is necessary to introduce a metric on the vector space $\mathcal{V}$. This metric corresponds to the characteristic impedance of the scattering decomposition and it is symmetric. Once a symmetric metric Z on $\mathcal{V}$ has been introduced, it is possible to define a $(2,0)$ tensor:

$$Y^{li} = \bar{B} \begin{pmatrix} Z^{-1} & 0 \\ 0 & Z \end{pmatrix} \bar{B}^T \tag{1.67}$$

Making the tensorial product between tensors T and Y, it is possible to obtain a $(1,1)$ tensor as illustrated in the following equation where Einstein's notation (i.e. summation over repeated indexes) has been adopted:

$$L_j^l = Y^{li} T_{ij} = \bar{B} \begin{pmatrix} 0 & Z^{-1} \\ Z & 0 \end{pmatrix} \bar{B}_*^T \tag{1.68}$$

Since L is a $(1,1)$ tensor, it does make sense to talk about its eigenvalues [72]. Consider a numerical expression with respect to the chosen basis of an elements of $\mathcal{V} \times \mathcal{V}^*$. λ is an eigenvalue of L if and only if there exists an element (f, e) different from zero such that:

$$\lambda \begin{pmatrix} f \\ e \end{pmatrix} = \begin{pmatrix} 0 & Z^{-1} \\ Z & 0 \end{pmatrix} \begin{pmatrix} f \\ e \end{pmatrix} \tag{1.69}$$

Notice that the basis matrices disappeared in Eq.(1.69) because a numerical expression of elements of $\mathcal{V} \times \mathcal{V}^*$ has been used.
If λ is an eigenvalue of L it directly follows from Eq.(1.69) that:

$$\lambda f = Z^{-1} e \tag{1.70a}$$

$$\lambda e = Z f \tag{1.70b}$$

but, substituting Eq.(1.70a) in Eq.(1.70b) and Eq.(1.70b) in Eq.(1.70a) it directly follows that:

$$\begin{aligned} \lambda^2 f &= Z^{-1} Z f = f \\ \lambda^2 e &= Z Z^{-1} e = e \end{aligned} \tag{1.71}$$

Thus, Eq.(1.71) implies that the eigenvalues of the tensor L are $\lambda = +1$ and $\lambda = -1$, both with multiplicity $n(= \dim \mathcal{V})$. It is then possible to calculate two n-dimensional eigenspaces, corresponding to the eigenvalues 1 and -1 respectively, which depend on the choice of the metric Z on $\mathcal{V}$. This is indicated by:

$$\mathcal{V} \times \mathcal{V}^* = \mathcal{S}_Z^+ \oplus \mathcal{S}_Z^- \tag{1.72}$$

where $\oplus$ denotes the direct sum operator. This decomposition implies that, for each Z, there is a unique way to decompose a pair of power conjugated variables (f, e) in the sum of two elements $s^+ \in \mathcal{S}_Z^+$ and $s^- \in \mathcal{S}_Z^-$. The following linear algebra result holds [117]:

Theorem 1.31. *Given any symmetric, positive semidefinite matrix Z, there exists always a symmetric matrix N such that*

$$Z = N^T N = NN = N^2$$

N is called the symmetric square root of Z

Thus it is always possible to express the matrix representation of the metric as the square of a symmetric matrix.
It is possible to give a representation of the eigenspaces corresponding to the eigenvalues ± 1 both in kernel form and in image form:

$$\begin{aligned} \mathcal{S}_Z^+ &= ker\left(I \; -Z^{-1}\right) \bar{B}_*^T = Im\bar{B} \begin{pmatrix} Z^{-1} \\ I \end{pmatrix} \frac{N}{\sqrt{2}} \\ \mathcal{S}_Z^- &= ker\left(Z \; I\right) \bar{B}_*^T = Im\bar{B} \begin{pmatrix} -I \\ Z \end{pmatrix} \frac{N^{-1}}{\sqrt{2}} \end{aligned} \tag{1.73}$$

where N is the symmetric square root of Z.

Proposition 1.32. *The eigenspaces $\mathcal{S}_Z^+$ and $\mathcal{S}_Z^-$ are orthogonal with respect to the + pairing defined in Eq.(1.61)*

Proof. $\mathcal{S}_Z^+$ and $\mathcal{S}_Z^-$ are orthogonal if and only if their image representations reported in Eq.(1.73) are orthogonal with respect to the representation of the + pairing given by Eq.(1.66). Thus the following matrix:

$$\frac{N^{-T}}{\sqrt{2}}\left(-I\ Z^T\right)\bar{B}^T\begin{pmatrix}0 & I\\ I & 0\end{pmatrix}\bar{B}\begin{pmatrix}Z^{-1}\\ I\end{pmatrix}\frac{N}{\sqrt{2}} \tag{1.74}$$

should be zero. Recalling the definition of the basis matrix $\bar{B}$, by simple calculations it is possible to show that Eq.(1.74) can be reduced to:

$$\frac{N^{-T}}{2}\left(-I\ Z^T\right)\begin{pmatrix}0 & I\\ I & 0\end{pmatrix}\left(Z^{-1}\ I\right)N \tag{1.75}$$

which is equal to:

$$\frac{1}{2}N^{-T}(Z^TZ^{-1}-I)N \tag{1.76}$$

Since Z has been chosen symmetric, $Z^T = Z$ and the expression Eq.(1.76) is equal to zero.

The restriction of the $+$ pairing $\langle,\rangle_+$ on $\mathcal{V}\times\mathcal{V}^*$ to the eigenspace $\mathcal{S}_Z^+$ defines an inner product on $\mathcal{S}_Z^+$. Analogously, the restriction of $-\langle,\rangle_+$ to the eigenspace $\mathcal{S}_Z^-$ defines an inner product on $\mathcal{S}_Z^-$. It is possible to choose as basis for $\mathcal{S}_Z^+$ and $\mathcal{S}_Z^-$ the columns of

$$\bar{B}\begin{pmatrix}Z^{-1}\\ I\end{pmatrix}\frac{N}{\sqrt{2}} \tag{1.77}$$

and

$$\bar{B}\begin{pmatrix}-I\\ Z\end{pmatrix}\frac{N^{-1}}{\sqrt{2}} \tag{1.78}$$

Proposition 1.33. *Consider as a basis of $\mathcal{S}_Z^+$ the columns of the matrix given in Eq.(1.77). This basis is orthonormal with respect to the inner product induced by the $+$ pairing on $\mathcal{S}_Z^+$.*
Consider as a basis of $\mathcal{S}_Z^-$ the columns of the matrix given in Eq.(1.78). This basis is orthonormal with respect to the inner product induced by the $+$ pairing on $\mathcal{S}_Z^-$.

Proof. Consider the basis defined by Eq.(1.77) for $\mathcal{S}_Z^+$. The following relation holds:

$$\begin{aligned}
&\tfrac{N^T}{\sqrt{2}}\left(Z^{-T}\ I\right)\bar{B}^T\begin{pmatrix}0 & I\\ I & 0\end{pmatrix}\bar{B}\begin{pmatrix}Z^{-1}\\ I\end{pmatrix}\tfrac{N}{\sqrt{2}} =\\
&\tfrac{N^T}{\sqrt{2}}\left(Z^{-T}\ I\right)\begin{pmatrix}0 & I\\ I & 0\end{pmatrix}\begin{pmatrix}Z^{-1}\\ I\end{pmatrix}\tfrac{N}{\sqrt{2}} =\\
&\frac{1}{2}N^T(Z^{-T}+Z^{-1})N = NZ^{-1}N = I
\end{aligned} \tag{1.79}$$

Thus the basis chosen for $\mathcal{S}_Z^+$ is orthonormal with respect to the induced inner product. Analogously, consider the basis defined in Eq.(1.78) for $\mathcal{S}_Z^-$. The following relation holds:

$$
\begin{aligned}
&-\tfrac{N^{-T}}{\sqrt{2}}\left(-I\ Z^T\right)\bar{B}^T\begin{pmatrix}0 & I\\ I & 0\end{pmatrix}\bar{B}\begin{pmatrix}-I\\ Z\end{pmatrix}\tfrac{N^{-1}}{\sqrt{2}} = \\
&-\tfrac{N^{-T}}{\sqrt{2}}\left(-I\ Z^T\right)\begin{pmatrix}0 & I\\ I & 0\end{pmatrix}\begin{pmatrix}-I\\ Z\end{pmatrix}\tfrac{N^{-1}}{\sqrt{2}} = \\
&\frac{1}{2}N^{-1}(Z+Z^T)N^{-1} = I
\end{aligned}
\tag{1.80}
$$

Thus the basis chosen for $\mathcal{S}_Z^-$ is orthonormal with respect to the induced inner product.

Remark 1.34. The proposition explains the presence of the factor $\sqrt{2}$ in the choice of the basis. This factor has the role of a normalization factor to obtain an orthonormal basis.

It is now possible to state the main result concerning scattering theory:

Theorem 1.35 (Scattering power decomposition). *Given any $(f,e) \in \mathcal{V}\times\mathcal{V}^*$ and any symmetric metric Z, the following relation holds:*

$$\langle e, f\rangle = \frac{1}{2}\|s_Z^+\|_+^2 - \frac{1}{2}\|s_Z^-\|_-^2 \tag{1.81}$$

where $s_Z^+ \in \mathcal{S}_Z^+$, $s_Z^- \in \mathcal{S}_Z^-$, $(f,e) = s_Z^+ + s_Z^-$ and $\|\cdot\|_+$ and $\|\cdot\|_-$ are the norms determined by the induced inner products on $\mathcal{S}_Z^+$ and $\mathcal{S}_Z^-$ respectively

Proof. Because of Eq.(1.72) it is possible to write, using the image representation of the eigenspaces:

$$\begin{pmatrix}f\\ e\end{pmatrix} = \frac{\bar{B}}{\sqrt{2}}\left(\begin{pmatrix}Z^{-1}\\ I\end{pmatrix}Ns_+ + \begin{pmatrix}-I\\ Z\end{pmatrix}N^{-1}s_-\right) \tag{1.82}$$

By straightforward calculation, it is possible to obtain:

$$\langle e, f\rangle = \frac{1}{2}(s_+^Ts_+ - s_-^Ts_- - s_+^Ts_- + s_-^Ts_+) = \frac{1}{2}(s_+^Ts_+ - s_-^Ts_-) \tag{1.83}$$

which proves the result using the results in Proposition 1.33

The elements of $\mathcal{S}_Z^+$ and $\mathcal{S}_Z^-$ are called scattering variables. The result of Theorem 1.35 is fundamental since it allows to algebraically decompose the power in an incoming power wave, which can be interpreted as supplied power, and an outgoing power wave, which can be interpreted as extracted power. The way the decomposition is done depends on the choice of the metric Z.

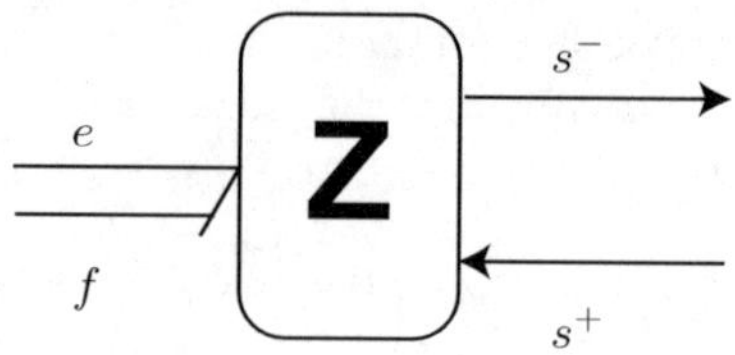

Fig. 1.8. Scattering power decomposition

A representation of the scattering power decomposition is given in Fig. 1.8; the half-arrow is associated to the pair $(e, f) \in \mathcal{V} \times \mathcal{V}^*$ and represents the power flow in bond-graph notation [237, 36]. It is possible to calculate the mapping between efforts flows and scattering variables in a given basis, i.e. with a numerical representation. In this case, using Eq.(1.82), it is possible to obtain:

$$\begin{cases} f = \frac{N^{-1}}{\sqrt{2}}(s_Z^+ - s_Z^-) \\ \\ e = \frac{N}{\sqrt{2}}(s_Z^+ + s_Z^-) \end{cases} \tag{1.84}$$

and

$$\begin{cases} s_Z^+ = \frac{N^{-1}}{\sqrt{2}}(e + Zf) \\ \\ s_Z^- = \frac{N^{-1}}{\sqrt{2}}(e - Zf) \end{cases} \tag{1.85}$$

Notice that the matrix $\bar{B}$ has now disappeared since numerical representation in the basis represented by $\bar{B}$ are considered.

The importance of the metric Z is twofold: geometric and physical. Since it does not make sense to talk about eigenvalues of a $(0, 2)$ tensor (see [72]), it is geometrically necessary to introduce a metric on $\mathcal{V}$ to be able to decompose $\mathcal{V} \times \mathcal{V}^*$ in eigenspaces in a coordinate free way. Furthermore, the metric Z has a very clear physical interpretation. While it was possible to talk about power without introducing any structure on $\mathcal{V}$, the introduction of the metric provide an extra information, namely the sense of propagation of the power, which is the basis of the scattering power decomposition. In fact, without a metric, it wouldn't be possible to talk about an *incoming* and an *outgoing* power wave.

Scattering variables will play a key role when considering communication delays in interactive robotic system as it will be shown in Chap. 3 and in Chap. 4.

1.6 Conclusions

Energy exchange plays a key role both for the description and for the control of the physical interaction. The port-Hamiltonian framework is a very useful tool for tackling the problem of controlling interactive robotic interfaces. In fact, the port-Hamiltonian description of physical systems puts into evidence all the energetic properties of the system: the amount of energy stored, through the state energy variables, the energy dissipation, through the dissipative elements, the interfaces with the external world, through the power ports, and the interconnection structure along which the parts of the system exchange energy. All these features can be fruitfully exploited during the control design. Furthermore, the port-Hamiltonian formalism allows to model any lumped parameters physical system. Thus, using this framework for modeling and controlling interactive robotic systems, allows to describe and to design control strategies also for complex (i.e. nonlinear) interactive interfaces.

There is still a lot of ongoing research about port-Hamiltonian systems. Recently, the concepts of Dirac structure and of power port have been extended for modeling distributed parameters physical systems, see for example [193, 190, 324, 247, 325, 103, 174, 173, 176]. The problem of the discretization of distributed port-Hamiltonian systems is also receiving considerable attention because of its importance for simulation, see [104, 105] and references therein.

The port-Hamiltonian framework is being exploited in the description of complex physical systems, possibly consisting of subsystems from different physical domains. For example, in [79, 80] it is possible to find applications in the domain of thermodynamics and in [64] the port-Hamiltonian framework is used for modeling a chemical reactor. In [238, 18] port-Hamiltonian systems are exploited for modeling and controlling electrical machines and in [77, 78, 75] they are used to model the dynamics of walking machines.

The scattering framework presented in this section is a generalization of the work reported in [7, 215] where scalar quantities where considered and no geometry was present. In the framework presented in this section, it is possible to make the scattering decomposition of Eq.(1.72) both in case that power variables are scalar and that are more complex geometric entities (e.g. twists and wrenches [272, 290]). The geometric framework for scattering has been recently extended to infinite dimensional systems, see [179, 173]. For some comments on the role of non symmetric metrics in the scattering representation, see [299, 319].

2

Control of Port-Hamiltonian Systems

2.1 Introduction

Energy plays a central role in the control of physical systems since the "shape" of the energy is related to the stability properties of the system. In fact, it is well known from physics, that every configuration characterized by a (local) minimum of the energy exhibits a (locally) stable behavior. Unfortunately the configuration that naturally corresponds to a minimum of the energy is very seldom the desired configuration for the system.

One of the main strategies behind the control of physical systems is the so-called *energy shaping*. The controller is interpreted as a device that exchanges energy with the plant and it has to be designed in such a way that the controlled system can still be interpreted as a physical system which has an energy function whose minimum corresponds to the desired configuration for the plant.

The idea of energy shaping has been used for the first time in [306] for the control of a robotic manipulator and has led to the very well-know PD plus gravity compensation control. Energy shaping techniques have been fruitfully applied to systems described in the Euler-Lagrange formalism, for a reference see for example [226].

In Chap. 1 it has been shown that the port-Hamiltonian framework is very well suited for modeling physical systems and, therefore, it is possible to cast the problem of regulation of a physical system into the problem of regulation of a port-Hamiltonian system without loss of generality. The biggest advantage brought by the port-Hamiltonian formalism is that all the energetic properties (the stored energy but also the interconnection structure and the inherent dissipation) of the system are very evident and thus they can be very easily exploited for regulation purposes, shedding a new light on energy shaping techniques and allowing to develop new energy-based control algorithms.

In this chapter we introduce the main energy based control techniques for port-Hamiltonian systems. The link between stability and energetic properties of a physical system can be formalized by means of *passivity theory*. Thus, the basic concepts of passivity theory and the relation between port-Hamiltonian systems and passive systems are reported in Sec. 2.2. In Sec. 2.3 we present the energy shaping control strategy for port-Hamiltonian systems, following the geometric interpretation proposed in [188, 228, 186, 227, 319], and we show what are the limitations of this control strategy which are particularly evident when considering electrical machines. In Sec. 2.4 we show that the limitations of the energy shaping can be overcome by exploiting the port-Hamiltonian structure for assigning not only the shape of the energy function of the closed loop but also the interconnection structure and the inherent dissipation and we shortly introduce the Interconnection and Damping Assignment Passivity Based Control (IDA-PBC) strategy proposed in [227]. Finally, in Sec. 2.5, we show how it is possible to embed variable structure techniques into energy shaping, as proposed in [178, 177, 256], in order to achieve perfect regulation also in the case in which the physical parameters of the port-Hamiltonian plant are unknown.

2.2 Basic Concepts of Passivity Theory

2.2.1 Definitions and Properties

Consider an I/S/O dynamical system (see Def. 1.16) represented by the following nonlinear differential equations (*affine system*):

$$\begin{cases} \dot{x} = f(x) + g(x)u \\ \\ y = h(x) \end{cases} \tag{2.1}$$

where $x \in \mathcal{X}$ is the state variable and $\mathcal{X}$ is the state space, $u \in \mathcal{U}$ is the input, $y \in \mathcal{Y}$ is the output and $\mathcal{U}$ and $\mathcal{Y}$ are m dimensional input and output spaces respectively. Furthermore f and g are smooth vector fields and h is a smooth mapping. Assume that the system has at least one equilibrium configuration; without loss of generality, $x = 0$ can be taken as the equilibrium point. In the following, x_0 will denote the initial state and $t = 0$ will denote the initial instant of time. Furthermore, $\Phi(t, x_0, u)$ will denote the value of the state trajectory at time $t \geq 0$ when the system starts from an initial state $x(0) = x_0$ and the input u is given.
Let w be a real valued map defined on $\mathcal{U} \times \mathcal{Y}$

$$w : \mathcal{U} \times \mathcal{Y} \rightarrow \mathbb{R} \tag{2.2}$$

called *supply rate* and let $\mathbb{R}^+$ denote the set of positive real numbers.

Definition 2.1 (Dissipative system). *A system of the form Eq.(2.1) is said to be dissipative with respect to the supply rate w if there exists a continuous function $V : \mathcal{X} \rightarrow \mathbb{R}^+$, called* storage function, *such that for all $u \in \mathcal{U}$, $x_0 \in \mathcal{X}$ and $t \geq 0$, the following relation, called* dissipation inequality *holds:*

$$V(x(t)) - V(x_0) \leq \int_0^t w(\tau)d\tau \qquad (2.3)$$

An important concept related to dissipative systems is the notion of *available storage* which can be defined as follows:

Definition 2.2 (Available Storage). *The available storage, denoted with V_a of a system with supply rate w is the function $V_a : \mathcal{X} \to \mathbb{R}$ defined by:*

$$V_a(x) = \sup_{x_0=x,\, u\in\mathcal{U},\, t\geq 0} \left\{ -\int_0^t w(\tau)d\tau \right\} \qquad (2.4)$$

It is important to note that the available storage, when defined, is non negative since it is the supremum over a set of number that contains the zero element (obtained when $t = 0$). The available storage can be used to check whether a system is dissipative or not. The following result holds:

Proposition 2.3 (Willems 1972). *If a system is dissipative with respect to a supply rate w, the available storage $V_a(x)$ is finite for each $x \in \mathcal{X}$. Furthermore, any possible storage function V satisfies*

$$0 \leq V_a(x) \leq V(x) \qquad (2.5)$$

for each $x \in \mathcal{X}$ and if V_a is continuous, then V_a itself is a storage function. Conversely, if $V_a(x)$ is finite for each $x \in \mathcal{X}$ and it is continuous, then the system is dissipative.

It is possible to give an energetic interpretation of dissipative systems. In fact the supply rate w and the storage function V can be thought as generalized power and a generalized energy respectively. The dissipation inequality expresses the fact that a system is dissipative if and only if the stored generalized energy at time t, $V(x(t))$, is at most equal to the sum of the initially stored generalized energy $V(x_0)$ and the total externally supplied generalized energy $\int_0^t w(u(\tau), y(\tau))d\tau$ in the interval $[0, t]$. This means that there cannot be any internal production of generalized energy but only dissipation is possible. The pair (u, y) represents, through the supply rate function, the medium through which the system can exchange generalized energy. Furthermore, since the storage function is nonnegative, the following inequality directly follows from Eq.(2.3):

$$-\int_0^t w(u(\tau), y(\tau))d\tau \leq V(x_0) \leq \infty \qquad (2.6)$$

which expresses the fact that the total amount of generalized energy that can be extracted from a dissipative system is bounded by the amount that is initially stored.

An important class of dissipative systems is that for which the generalized power coincides with power, generalized energy coincides with energy and the medium through which the system exchanges energy is a power port. Thus, an important choice of the supply rate can be made when $\mathcal{U}$ and $\mathcal{Y}$ are dual spaces and it is given by:

$$w(u(\tau), y(\tau)) = \langle u(\tau), y(\tau)\rangle \quad u \in \mathcal{U} \quad y \in \mathcal{Y} = \mathcal{U}^* \tag{2.7}$$

In this case the supply rate represents a power flow, u and y a flow and effort pair and $(\mathcal{U} \times \mathcal{Y}, w)$ a *power port.* The storage function V represents the energy stored into the system.

It is possible to represent the m dimensional input and output spaces $\mathcal{U}$ and $\mathcal{Y} = \mathcal{U}^*$ with $\mathbb{R}^m$. Within this representation both $u \in \mathcal{U}$ and $y \in \mathcal{Y}$ are represented by m dimensional column vectors and the intrinsically defined duality product between $\mathcal{U}$ and $\mathcal{Y}$ is expressed as follows:

$$\langle u, y\rangle = y^T u \tag{2.8}$$

Definition 2.4 (Passive system). *A system is* passive *if it is dissipative with respect to the supply rate $w(u, y) = \langle u, y\rangle = y^T u$.*

For any passive system the dissipation inequality turns out to be:

$$V(x(t)) - V(x_0) \leq \int_0^t y^T(\tau)u(\tau)d\tau \tag{2.9}$$

Two important properties follow from the fact that the storage function is decreasing along the trajectories. In fact, if $u = 0$ then:

$$\dot{V}(x(t)) \leq y^T(t)u(t) \leq 0 \qquad \forall t \tag{2.10}$$

Thus, if $V(x)$ is positive definite, the equilibrium point $x = 0$ is Lyapunov stable. Furthermore, if $y = 0$ Eq.(2.10) keeps on holding. Thus, if V is positive definite, the *zero dynamics* [128] of the system is Lyapunov stable.

Remark 2.5. It can be easily shown that any strict local minimum x_m of the storage function is Lyapunov stable by considering as Lyapunov function $V(x) - V(x_m)$. The proof is immediate by considering Eq.(2.9).

It is often useful to distinguish passive systems for which the dissipation inequality becomes either an equality or a strict inequality.

Definition 2.6 (Lossless system). *A passive system with storage function V is* lossless *if for all $u \in \mathcal{U}$ and for all $x_0 \in \mathcal{X}$ and $t \geq 0$*

$$V(x(t)) - V(x_0) = \int_0^t y^T(\tau)u(\tau)d\tau \tag{2.11}$$

Definition 2.7 (Strictly passive system). *A passive system with storage function V is* strictly passive *if there exists a positive definite function $S : \mathcal{X} \to \mathbb{R}^+$ such that for all $u \in \mathcal{U}$, $x_0 \in \mathcal{X}$ and $t \geq 0$:*

$$V(x(t)) - V(x_0) = \int_0^t y^T(\tau)u(\tau)d\tau - \int_0^t S(x(\tau))d\tau \qquad (2.12)$$

Remark 2.8. A lossless system stores all the energy provided through the power port while a strictly passive system dissipates part of it and the amount of energy that is dissipated is given by $\int_0^t S(x(\tau))d\tau$.

In the following we will indicate with $L_f V(x)$ and with $L_g V(x)$ the Lie derivative of $V(x)$ with respect to the vector fields f and g respectively, see Sec. A.1.
It is possible to establish a link between passive systems and the nonlinear version of Kalman-Yakubovitch-Popov (KYP) lemma which will be useful for relating passivity and port-Hamiltonian systems.

Definition 2.9 (KYP property). *A non linear system described by Eq.(2.1) enjoys the KYP property if there exists a non negative $\mathcal{C}^1$ function $V : \mathcal{X} \to \mathbb{R}$ with $V(0) = 0$ such that:*

$$L_f V(x) \leq 0 \qquad (2.13a)$$

$$L_g V(x) = h^T(x) \qquad (2.13b)$$

for each $x \in \mathcal{X}$.

The following proposition holds:

Proposition 2.10 (Byrnes *et al.* 1991). *If a system enjoys the KYP property then it is passive. Conversely, if a system is passive with a $\mathcal{C}^1$ storage function then it enjoys the KYP property.*

Proof. If the system enjoys the KYP property then:

$$\begin{aligned}\frac{dV}{dt} = \frac{\partial V}{\partial x}\dot{x} = \frac{\partial V}{\partial x}f(x(t)) + \frac{\partial V}{\partial x}g(x(t))u(t) = L_f V(x(t)) + L_g V(x(t))u(t) \leq \\ \leq L_g V(x(t))u(t) = y^T(t)u(t)\end{aligned} \qquad (2.14)$$

Thus the system is passive with $V(x)$ as storage function.
Conversely, if a system is passive with a $\mathcal{C}^1$ storage function, the following inequalities holds:

$$\frac{dV}{dt} = \frac{\partial V}{\partial x}f(x) + \frac{\partial V}{\partial x}g(x)u \leq y^T(t)u(t) \qquad (2.15)$$

which clearly implies Eq.(2.13a) and Eq.(2.13b)

It directly follows from definitions that for a lossless system $L_f V(x(t)) = 0$ and that for a strictly passive system $L_f V(x(t)) = -S(x)$. Thus, if a system is strictly passive and its storage function is positive definite in its equilibrium point $x = 0$, then the equilibrium point is Lyapunov stable.

The term $L_f V(x(t))$ represents the dissipation present in the system and the term $L_g V(x(t))u(t) = y^T(t)u(t)$ represents the power flowing through the power port. In fact, in a lossless system $L_f V(x(t)) = 0$, namely there is not dissipation and all the energy supplied is stored. It is then possible to exploit the KYP property to give an alternative differential definition of a passive system:

Definition 2.11. *A system is passive if the power supplied is either stored or dissipated, i.e. if:*

$$P = \frac{dV}{dt} + P_{diss} \tag{2.16}$$

where $V(x)$ is the storage function, $P = L_g V(x(t))u = y^T u$ is the supplied power and $P_{diss} = -L_f V(x(t)) > 0$ is the power dissipated by the system.

For lossless systems all the power supplied is stored while for strictly dissipative systems a portion of the supplied power is dissipated while the rest is stored. If the relation Eq.(2.16) is satisfied with $P_{diss} < 0$ it means that the system is not passive since a negative dissipation represents an extra power injection into the system.

2.2.2 Output Feedback Stabilization of Passive Systems

Passive systems have been and are very much studied because of the appealing link between passivity and Lyapunov stability, shown in Sec. 2.2.1. The aim of this subsection is to show how it is possible to *asymptotically* stabilize a passive system in its equilibrium point (which is again supposed to exist and to be $x = 0$) using very well established techniques [41, 129, 319, 170].

Definition 2.12 (zero state detectability and observability). *A system described by Eq.(2.1) is* locally zero state detectable *if there exists a neighborhood $U \subset \mathcal{X}$ of 0 such that, for all $x \in U$*

$$\text{If} \quad h(\Phi(t, x, 0)) = 0 \quad \forall t \geq 0 \quad \text{then} \quad \lim_{t \to \infty} \Phi(t, x, 0) = 0 \tag{2.17}$$

If $U = \mathcal{X}$ then the system is said zero state detectable.
A system is locally zero state observable *if there exists a neighborhood $U \subset \mathcal{X}$ of 0 such that, for all $x \in U$*

$$\text{If} \quad h(\Phi(t, x, 0)) = 0 \quad \forall t \geq 0 \quad \text{then} \quad x = 0 \tag{2.18}$$

If $U = \mathcal{X}$ the the system is said zero state observable

Definition 2.13 (proper function). *A nonnegative function* $V : \mathcal{X} \to \mathbb{R}$ *is* proper *if for each* $a > 0$ *the set*

$$V^{-1}([0,a]) = \{x \in \mathcal{X} \mid 0 \leq V(x) \leq a\} \tag{2.19}$$

is compact.

Remark 2.14. Proper functions are sometimes called radially unbounded functions [143].

It is now possible to prove the following theorem [41]:

Theorem 2.15 (Output feedback asymptotic stabilization). *Consider a system described by Eq.(2.1) that is passive with a positive definite storage function* V*, that is locally zero state detectable and that admits* $x = 0$ *as an equilibrium configuration. Let* $\phi : \mathcal{Y} \to \mathcal{U}$ *be a smooth function such that* $\phi(0) = 0$ *and* $y^T\phi(y) > 0$ *for each non zero* y*. The control law:*

$$u = -\phi(y) \tag{2.20}$$

asymptotically stabilizes the equilibrium point. If the system is zero state detectable and V *is proper, then the control law given by Eq.(2.20) globally asymptotically stabilizes the system.*

Proof. Since the system is passive the dissipation inequality holds and therefore, considering $u = -\phi(y)$:

$$V(x(t)) - V(x_0) \leq -\int_0^t y^T(\tau)\phi(y(\tau))d\tau \leq 0 \tag{2.21}$$

Thus the storage function is non decreasing along the trajectories of the closed loop system. Since V is positive definite, the equilibrium point $x = 0$ is Lyapunov stable. Choose a sufficiently small neighborhood of the equilibrium (i.e. such that the stability property holds) and consider an initial condition x^0 in that neighborhood. Denote with $x^0(t)$ and γ^0 the state trajectory corresponding to x^0 and the correspondent positive limit set (which is non-empty, compact and invariant). Since $\lim_{t\to\infty} V(x^0(t)) = a_0 \geq 0$, by the continuity of V follows that $V(x) = a_0$ for all $x \in \gamma^0$. Let $\bar{x} \in \gamma^0$ and $\bar{x}(t)$ the corresponding trajectory. Since $\bar{x}(t) \in \gamma^0$ for all $t \geq 0$ then

$$0 = V(\bar{x}(t)) - V(\bar{x}) \leq -\int_0^t y^T(\tau)\phi(y(\tau))d\tau \leq 0 \tag{2.22}$$

implies $y = 0$ for all $t \geq 0$. By detectability $\lim_{t\to\infty} \bar{x}(t) = 0$ and therefore $a_0 = 0$. Thus $\lim_{t\to\infty} V(x^0(t)) = 0$, i.e. $\lim_{t\to\infty} x^0(t) = 0$. This proves local asymptotic stability of the equilibrium point $x = 0$. If V is proper and the system is zero state detectable, then the equilibrium point is globally asymptotically stable.

Remark 2.16. It can be shown by a change of coordinate that any strict minimum of the storage function can be (locally) asymptotically stabilized by static output feedback.

There are many corollaries of Theorem 2.15. One of the most important is the following:

Corollary 2.17. *Suppose that a system, which admits $x = 0$ as an equilibrium configuration, is lossless with an at least $\mathcal{C}^1$ proper positive definite storage function V. If the system is zero state observable then for each $k > 0$ the control law $u = -ky$ globally asymptotically stabilizes the equilibrium point $x = 0$.*

A lot of results on stabilization of nonlinear systems can be reinterpreted in the light of passivity theory and of Theorem 2.15. For example, a very well known result about the stabilization of affine systems obtained in [137] using nonlinear geometric control techniques states that if there exists a function $V(x)$ with only one minimum point in $x = 0$ and such that

$$L_f V(x(t)) = 0 \tag{2.23}$$

then the control law

$$u = -(L_g V(x(t)))^T \tag{2.24}$$

globally asymptotically stabilizes the equilibrium point. This is equivalent to state that if the system is lossless with storage function V it is possible to apply Corollary 2.17 with $k = 1$ to stabilize the equilibrium point $x = 0$.

2.2.3 Port-Hamiltonian Systems and Passivity

It has already been pointed out from an energetic point of view that the behavior of a port-Hamiltonian system is either lossless (Eq.(1.31)), in case no dissipation is present in the system, or passive, in case some dissipation is present into the system (Eq.(1.33)).

In this section a formal proof of the passivity properties of port-Hamiltonian system using a coordinate-based approach will be given and, for the sake of clearness, explicit port-Hamiltonian systems will be considered; the obtained results can be easily extended to implicit port-Hamiltonian systems. Consider a port-Hamiltonian system with dissipation:

$$\begin{cases} \dot{x} = (J(x) - R(x))\dfrac{\partial H}{\partial x} + g(x)u \\ \\ y = g^T(x)\dfrac{\partial H}{\partial x} \end{cases} \tag{2.25}$$

where $x \in \mathcal{X}$ is the state of the system and $H : \mathcal{X} \to \mathbb{R}$ is the Hamiltonian function which represents the energy stored and is non negative. The input signal space $\mathcal{U}$ and the output signal spaces $\mathcal{Y}$ are dual m-dimensional vector spaces, i.e. $\mathcal{Y} = \mathcal{U}^*$. Furthermore, $J(x)$ is a skew symmetric matrix representing the internal power preserving interconnections, $R(x)$ is a symmetric positive semidefinite matrix representing the dissipation of the system and $g(x)$ is a matrix describing the way that power coming from the external world is distributed into the system.
The link between passivity and port-Hamiltonian systems is stated in the following:

Proposition 2.18. *A port-Hamiltonian system with dissipation is a passive system and the storage function is the Hamiltonian function.*

Proof. A port-Hamiltonian system with dissipation (Eq.(2.25))can be interpreted as an affine system (Eq.(2.1)) where:

- $f(x) = (J(x) - R(x))\frac{\partial H}{\partial x}$
- $g(x) = g(x)$
- $h(x) = g^T(x)\frac{\partial H}{\partial x}$

The following relation holds:

$$L_{(J(x)-R(x))\frac{\partial H}{\partial x}} H(x) = \frac{\partial^T H}{\partial x}(J(x) - R(x))\frac{\partial H}{\partial x} = -\frac{\partial^T H}{\partial x}R(x)\frac{\partial H}{\partial x} \leq 0 \tag{2.26}$$

where the skew symmetry of $J(x)$ has been exploited and the inequality follows from the fact that $R(x)$ is positive semidefinite. Furthermore:

$$L_g H(x) = \frac{\partial^T H}{\partial x} g(x) = (g^T(x)\frac{\partial H}{\partial x})^T \tag{2.27}$$

Thus a port-Hamiltonian system with dissipation enjoys the KYP property and consequently, applying Proposition 2.10, a port-Hamiltonian system with dissipation is a passive system.

If $R(x) = 0$, namely if there is no dissipation in the system, then

$$L_{(J(x)-R(x))\frac{\partial H}{\partial x}} H(x) = 0 \tag{2.28}$$

Therefore, a port-Hamiltonian system without dissipation is a lossless system. Furthermore if $R(x)$ is positive definite, the port-Hamiltonian system is strictly passive. Thus, losslessness, passivity and strict passivity of a port-Hamiltonian system can be determined by simply checking the sign of the matrix $R(x)$. This can be very clearly interpreted using Def. 2.11. In fact, by straightforward calculations, the following relation can be obtained:

$$P = y^T u = \frac{dH}{dt} + \underbrace{\frac{\partial^T H}{\partial x} R(x) \frac{\partial H}{\partial x}}_{P_{diss}} \tag{2.29}$$

The sign of the dissipated power depends on $R(x)$. If $R(x)$ is positive definite, $P_{diss} > 0$ which means that some power is always dissipated by the system and that consequently the system is strictly passive. If $R(x) = 0$ then $P_{diss} = 0$ which means that there is no dissipation and that consequently the system is lossless. If $R(x)$ were negative definite, $P_{diss} < 0$ which means that the system is not passive since there is some internal production of energy.

Remark 2.19. In general, port-Hamiltonian systems are characterized by a lower bounded, non necessarily non negative, Hamiltonian function, namely we have that there is a finite positive constant $\zeta \in \mathbb{R}^+$ such that:

$$H(x) \geq -\zeta$$

In this case it is still possible to prove that a port-Hamiltonian system is a passive system by considering $H^*(x) = H(x) + \zeta$ as a storage function.

Thus, port-Hamiltonian systems inherit all the properties of passive systems. It is then possible to asymptotically stabilize an equilibrium configuration corresponding to a (local) minimum point of the Hamiltonian function by the control law $u = -ky$. This kind of control is called *stabilization by damping injection*. The name follows from the fact that the control action can be physically interpreted as the addition of some damping to the plant. In fact, consider a port-Hamiltonian system with dissipation where $u = -ky$ and $k > 0$. The controlled system is represented by the following equations:

$$\dot{x} = [(J(x) - R(x)] \frac{\partial H}{\partial x} - kg(x)g^T(x)\frac{\partial H}{\partial x} = \left[J(x) - (R(x) + kg(x)g^T(x))\right] \frac{\partial H}{\partial x} \tag{2.30}$$

The damping injection adds to the system some extra power dissipation which is modeled by the symmetric positive semidefinite matrix $kg(x)g^T(x)$.

Remark 2.20. In presence of damping, the state of the system evolves towards a configuration corresponding to a minimum of the Hamiltonian function. The rate of convergence is determined by the amount of energy that is extracted from the system. Thus, introducing further dissipation into the system (e.g. by damping injection) allows to increase the rate by which the system evolves towards a minimum energy configuration.

Remark 2.21. Recalling Remark 2.5 and Remark 2.16, it can be deduced that any strict minimum of the Hamiltonian function corresponds to a Lyapunov stable configuration that can be asymptotically stabilized by damping injection.

Example 2.22 (Linear Oscillator). Consider the linear oscillator composed by a mass and a linear spring represented in Fig. 1.3. The port-Hamiltonian model of the system is:

$$\begin{aligned} \begin{pmatrix} \dot{x} \\ \dot{p} \end{pmatrix} &= \begin{pmatrix} 0 & 1 \\ -1 & 0 \end{pmatrix} \begin{pmatrix} \frac{\partial H}{\partial x} \\ \frac{\partial H}{\partial p} \end{pmatrix} + \begin{pmatrix} 0 \\ 1 \end{pmatrix} u \\ y &= (0\ 1) \begin{pmatrix} \frac{\partial H}{\partial x} \\ \frac{\partial H}{\partial p} \end{pmatrix} \end{aligned} \tag{2.31}$$

where x and p are energy variables that denote the elongation of the spring and the momentum of the mass respectively. The input u is the force that acts on the mass and the output y is the velocity of the mass. The Hamiltonian function is the sum of the kinetic energy stored by the mass and of the potential (elastic) energy stored by the spring and has the following expression:

$$H = \frac{p^2}{2m} + \frac{1}{2}kx^2 \tag{2.32}$$

where m is the inertia of the mass and k is the stiffness of the spring.
It is easy to see that the point $(0, 0)$ is an equilibrium point and, thus, that it is a global minimum point of the Hamiltonian function. It is possible to make a stability analysis of the equilibrium point. Taking as candidate Lyapunov function the Hamiltonian of the system, the following relations hold:

$$\begin{aligned} &H(x,p) > 0 \quad \forall x \neq 0, \forall p \neq 0 \quad H(0,0) = 0 \\ &\frac{dH}{dt} = \begin{pmatrix} \frac{\partial^T H}{\partial x} & \frac{\partial^T H}{\partial p} \end{pmatrix} \begin{pmatrix} 0 & 1 \\ -1 & 0 \end{pmatrix} \begin{pmatrix} \frac{\partial H}{\partial x} \\ \frac{\partial H}{\partial p} \end{pmatrix} = 0 \end{aligned} \tag{2.33}$$

Thus, it follows that the equilibrium point is Lyapunov stable but NOT asymptotically stable. It is possible to asymptotically stabilize the equilibrium point by damping injection. Consider the following control law:

$$u = -ky = -k\,(0\ 1) \begin{pmatrix} \frac{\partial H}{\partial x} \\ \frac{\partial H}{\partial p} \end{pmatrix} \tag{2.34}$$

The controlled system is still a port-Hamiltonian system and is represented by the following equations:

$$\begin{aligned} \begin{pmatrix} \dot{x} \\ \dot{p} \end{pmatrix} &= \left[\begin{pmatrix} 0 & 1 \\ -1 & 0 \end{pmatrix} - k \begin{pmatrix} 0 & 0 \\ 0 & 1 \end{pmatrix} \right] \begin{pmatrix} \frac{\partial H}{\partial x} \\ \frac{\partial H}{\partial p} \end{pmatrix} + \begin{pmatrix} 0 \\ 1 \end{pmatrix} u \\ y &= (0\ 1) \begin{pmatrix} \frac{\partial H}{\partial x} \\ \frac{\partial H}{\partial p} \end{pmatrix} \end{aligned} \tag{2.35}$$

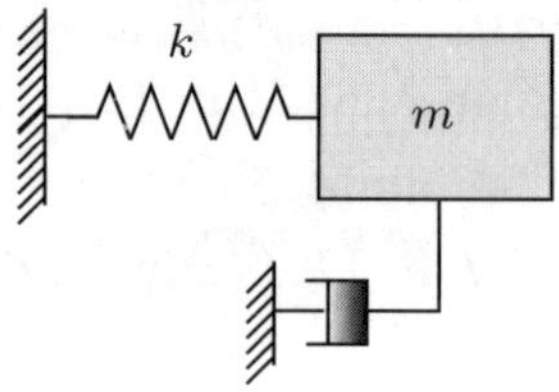

Fig. 2.1. The physical equivalent of the oscillator controlled by damping injection

The state $(0,0)$ is still an equilibrium point. Consider as a candidate Lyapunov function for the controlled system the Hamiltonian function; the following relations hold:

$$H(x,p) > 0 \quad \forall x \neq 0, \forall p \neq 0 \quad H(0,0) = 0$$

$$\frac{dH}{dt} = \begin{pmatrix} \frac{\partial^T H}{\partial x} & \frac{\partial^T H}{\partial p} \end{pmatrix} \begin{pmatrix} 0 & 1 \\ -1 & 0 \end{pmatrix} \begin{pmatrix} \frac{\partial H}{\partial x} \\ \frac{\partial H}{\partial p} \end{pmatrix} - \begin{pmatrix} \frac{\partial^T H}{\partial x} & \frac{\partial^T H}{\partial p} \end{pmatrix} \begin{pmatrix} 0 & 0 \\ 0 & 1 \end{pmatrix} \begin{pmatrix} \frac{\partial H}{\partial x} \\ \frac{\partial H}{\partial p} \end{pmatrix} = \tag{2.36}$$

$$= - \begin{pmatrix} \frac{\partial^T H}{\partial x} & \frac{\partial^T H}{\partial p} \end{pmatrix} \begin{pmatrix} 0 & 0 \\ 0 & 1 \end{pmatrix} \begin{pmatrix} \frac{\partial H}{\partial x} \\ \frac{\partial H}{\partial p} \end{pmatrix} \leq 0$$

Thus the Lyapunov function is negative semidefinite. But the set where the time derivative of the Hamiltonian is equal to zero is:

$$Z = \{(x,p) \mid p = 0\} \tag{2.37}$$

and the biggest invariant subset of Z is $\{(0,0)\}$. Therefore, by LaSalle's invariance principle [143], the equilibrium point is asymptotically stable.

The control by damping injection has a nice physical interpretation. In fact, since the main aim of the damping injection is to introduce some dissipation into the system, the controller can be interpreted as a "virtual" damper that is added to the mass that is composing the oscillator, as represented in Fig. 2.1.

2.3 Energy Shaping of Port-Hamiltonian Systems

In Sec. 2.2.3 it has been shown that the strict minima of the energy function (i.e. of the Hamiltonian function) correspond to Lyapunov stable equilibrium configurations that can be asymptotically stabilized via damping injection.

Unfortunately it is very often required to stabilize a port-Hamiltonian system in a configuration that does not correspond to a strict minimum of the energy function. Therefore it is necessary to introduce a controller whose task is to change the shape of the energy function of the controlled system in order to have a strict minimum in the configuration of interest. It is then possible to asymptotically stabilize the new minimum energy configuration by means of damping injection. This control strategy is the so-called *energy shaping + damping injection* perspective of control. Loosely speaking, it consists of two steps:

1. ***Energy Shaping***: Shape the energy of the plant by means of a proper control law in order to assign a strict minimum in the desired configuration.
2. ***Damping Injection***: Add dissipation to the system via damping injection in order to asymptotically stabilize the desired configuration

Consider the port-Hamiltonian system represented by Eq.(2.25). The following energy balance follows by the integration of Eq.(2.29):

$$H(x(t)) - H(x(0)) = \int_0^t u^T(\tau)y(\tau)d\tau - \int_0^t \frac{\partial^T H}{\partial x} R(x) \frac{\partial H}{\partial x} d\tau = \\ = \int_0^t u^T(\tau)y(\tau)d\tau - d(t) \tag{2.38}$$

where x is the state of the system and $H(x)$ is the total energy function. Input and output are power conjugated variables and $d(t)$ is a non negative function that represents the natural dissipation that is present into the system.

The control problem of energy shaping plus damping injection can be formalized in terms of state feedback as follows:
Let x^* be a desired equilibrium configuration. Select a control action $u = \beta(x) + v$ such that the closed loop dynamics satisfies the new energy balance equation:

$$H_d(x(t)) - H_d(x(0)) = \int_0^t v^T(\tau)z(\tau)d\tau - d_d(t) \tag{2.39}$$

where H_d is the desired energy function with a strict minimum in x_* and z (which may be equal to y) is the new power conjugated output. Furthermore $d_d(t)$ is the desired dissipation of the closed loop system which can be assigned by damping injection.

2.3.1 Stabilization by Energy Balancing

There is a wide class of systems (which includes mechanical systems) for which the solution of the energy shaping problem is quite simple, [227].

Proposition 2.23. *If it is possible to find a function $\beta(x)$ such that:*

$$-\int_0^t \beta^T(x(\tau))y(\tau)d\tau = H_a(x(t)) + k \tag{2.40}$$

where k is a positive constant, then the control law $u = \beta(x) + v$ is such that the energy balance

$$H_d(x(t)) - H_d(x(0)) = \int_0^t v^T(\tau)y(\tau)d\tau - d(t) \tag{2.41}$$

is satisfied with $H_d(x) = H(x) + H_a(x)$

Proof. By replacing the control law $u = \beta(x) + v$ into Eq.(2.38), we get:

$$H(x(t)) - H(x(0)) = \int_0^t \beta^T(x(\tau))y(\tau)d\tau + \int_0^t v^T(\tau)y(\tau)d\tau - d(t) \tag{2.42}$$

Substituting Eq.(2.40) into Eq.(2.42), the following equation holds:

$$H(x(t)) - H(x(0)) = -H_a(x(t)) - k + \int_0^t v^T(\tau)y(\tau)d\tau - d(t) \tag{2.43}$$

from which it follows that:

$$H(x(t)) + H_a(x(t)) - H(x(0)) + k = \int_0^t v^T(\tau)y(\tau)d\tau + d(t) \tag{2.44}$$

From Eq.(2.40) it follows that necessarily $H_a(x(0)) = -k$ and thus Eq.(2.44) can be rewritten as:

$$H(x(t)) + H_a(x(t)) - H(x(0)) - H_a(x(0)) = \int_0^t v^T(\tau)y(\tau)d\tau + d(t) \tag{2.45}$$

Finally, setting $H_d(x(t)) = H(x(t)) + H_a(x(t))$ the balance of Eq.(2.41) follows.

Remark 2.24. The term $\int_0^t \beta^T(x(\tau))y(\tau)d\tau$ can be interpreted as the energy supplied by the controller to the plant. Therefore the condition of Eq.(2.40) expresses the fact that the energy supplied by the controller can be expressed as a function of the state.

The energy of the closed loop system is the difference between the energy stored by the plant and the energy supplied by the controller; therefore this energy shaping strategy is called *energy balancing passivity based control* (energy balancing PBC).

If the closed loop energy $H_d(x)$ has a strict minimum in the desired configuration x^*, then, setting $v = 0$, x^* is stable and the Lyapunov function is represented by the difference between the energy stored by the system and the energy supplied by the controller.

Example 2.25. Consider an n-DOF fully-actuated mechanical system with generalized coordinates $q \in \mathcal{Q}$. The port-Hamiltonian model of this system has been obtained in Example 1.30 and is:

$$\begin{cases} \begin{pmatrix} \dot{q} \\ \dot{p} \end{pmatrix} = \left[\begin{pmatrix} 0 & I_n \\ -I_n & 0 \end{pmatrix} - \begin{pmatrix} 0 & 0 \\ 0 & D(q,p) \end{pmatrix} \right] \begin{pmatrix} \frac{\partial H}{\partial q} \\ \frac{\partial H}{\partial p} \end{pmatrix} + \begin{pmatrix} 0 \\ B(q) \end{pmatrix} u \\ \\ y = \begin{pmatrix} 0 & B^T(q) \end{pmatrix} \begin{pmatrix} \frac{\partial H}{\partial q} \\ \frac{\partial H}{\partial p} \end{pmatrix} \end{cases} \tag{2.46}$$

with $H(q,p) = \frac{1}{2} p^T M^{-1}(q) p + V(q)$, where $V(q)$ is the potential energy, and where $D(q)$ is a positive definite matrix representing the viscous friction present in the system. Suppose that $q_d \in \mathcal{Q}$ is a desired configuration in the joint space. A possible function $\beta(\cdot)$ that satisfies Eq.(2.40) is:

$$\beta(q) = B^{-1}(q)[\frac{\partial V}{\partial q} - \frac{\partial H_a}{\partial q}] \tag{2.47}$$

where $H_a(q)$ is a function with a strict minimum in q_d. A simple choice of the function$\beta(\cdot)$ is given by:

$$\beta(q) = B^{-1}(q)[\frac{\partial V}{\partial q} - K_P(q - q_d)] \tag{2.48}$$

where $K_P = K_P^T$ is a positive definite proportional gain matrix. In fact, substituting Eq.(2.48) in Eq.(2.40) and recalling that

$$y = B^T(q)\frac{\partial H}{\partial p} = B^T(q)\dot{q}$$

the following relation follows:

$$- \int_0^t \beta^T(q(\tau))y(\tau)d\tau = - \int_0^t [\frac{\partial^T V}{\partial q} - (q-q_d)^T K_P] B^{-T}(q) B^T(q) \dot{q} d\tau =$$
$$-V(q(t)) + \frac{1}{2}(q-q_d)^T K_P (q-q_d) + k = H_a(q) + k \tag{2.49}$$

Thus the closed loop energy function is given by:

$$H_d(q,p) = H(q,p) + H_a(q) = \frac{1}{2} p^T M^{-1}(q) p + \frac{1}{2}(q-q_d)^T K_P (q-q_d) \tag{2.50}$$

which has a minimum in $(0, q_d)$ as desired. The point $(0, q_d)$ is asymptotically stable because of the presence of some inherent dissipation, modeled by the positive definite matrix $D(q)$, in the system. It is possible to add some further dissipation into the system by damping injection.

The main obstacle of the energy balancing control technique relies in finding $\beta(x)$ and $H_a(x)$ such that Eq.(2.40) is satisfied. In order to bring in evidence the main drawback of this control technique, it is more convenient to write Eq.(2.40) in its differential equivalent form, namely:

$$\dot{H}_a(x(t)) = -\beta^T(x(t))y(t) \tag{2.51}$$

Considering as a plant the port-Hamiltonian system represented by Eq.(2.25), Eq.(2.51) can be expanded as:

$$\begin{aligned}&\frac{\partial^T H_a}{\partial x}[(J(x(t)) - R(x(t)))\frac{\partial H}{\partial x} + g(x)\beta(x(t))] = -\beta^T(x(t))y(t) = \\ &= -\beta^T(x(t))g^T(x(t))\frac{\partial H}{\partial x}\end{aligned} \tag{2.52}$$

A necessary condition for the solvability of Eq.(2.52) is that $\beta^T(x(t))y(t)$ vanishes in correspondence of all the zeros of

$$(J(x(t)) - R(x(t)))\frac{\partial H}{\partial x} + g(x)\beta(x(t)) \tag{2.53}$$

Equilibrium configurations are obviously zeros of Eq.(2.53) and thus at every equilibrium point $\bar{x}$ it must be:

$$\beta^T(\bar{x})y(t) = 0 \tag{2.54}$$

Recalling that $-\beta^T(x(t))y(t)$ is the power extracted by the controller, the condition of Eq.(2.54) states that there must not be any power extraction at the equilibrium. This means that energy balancing PBC is applicable only if the energy dissipated by the system is bounded and, consequently, if it can be stabilized by extracting a finite amount of energy. This condition is satisfied for mechanical systems where the system has to be regulated at a configuration characterized by zero velocity. In the case of electrical circuits and electrical machines this is not always the case and thus energy-balancing PBC can fail. This drawback is known as the *dissipation obstacle.*

Example 2.26. Consider the series RLC circuit shown in Fig. 2.2. The energy variables are the charge q in the capacitor and the flux ϕ in the inductance; thus the state of the system can be chosen as $x = (q, \phi)^T = (x_1, x_2)^T$. The total energy function of the system is

$$H(x) = \frac{1}{2C}x_1^2 + \frac{1}{2L}x_2^2 \tag{2.55}$$

and the port-Hamiltonian model of the system is:

$$\begin{cases} \dot{x} = \begin{pmatrix} \dot{x}_1 \\ \dot{x}_2 \end{pmatrix} = \left[\begin{pmatrix} 0 & 1 \\ -1 & 0 \end{pmatrix} - \begin{pmatrix} 0 & 0 \\ 0 & R \end{pmatrix}\right] \begin{pmatrix} \frac{\partial H}{\partial x_1} \\ \frac{\partial H}{\partial x_2} \end{pmatrix} + \begin{pmatrix} 0 \\ 1 \end{pmatrix} u \\ \\ y = (0\ 1) \begin{pmatrix} \frac{\partial H}{\partial x_1} \\ \frac{\partial H}{\partial x_2} \end{pmatrix} \end{cases} \tag{2.56}$$

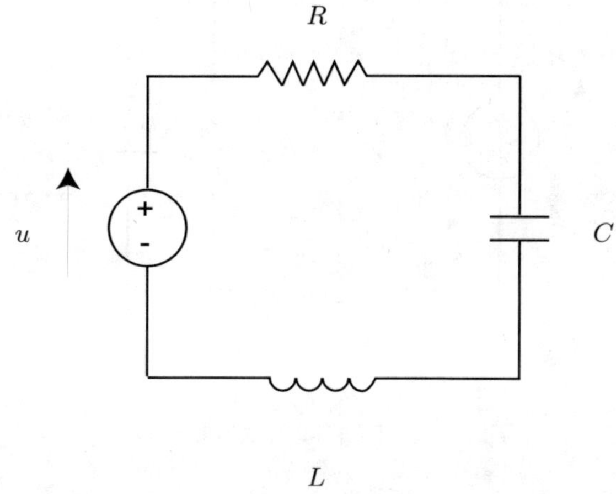

Fig. 2.2. Series RLC circuit

The system satisfies Eq.(2.38) with $d(t) = R\int_0^t[(1/L)x_2]^2(\tau)d\tau$, i.e. the energy dissipated by the resistor. It is clear from the constitutive equations that the admissible equilibrium configuration of the system are of the form $x^* = (x_1^*, 0)^T$. Notice that in correspondence of the equilibrium configuration there is no dissipation. In order to apply the energy-balancing PBC, Eq.(2.52) has to be solved. In this case $\beta(x)$ is a real valued function and the equation becomes:

$$\frac{1}{L}x_2\frac{\partial H_a}{\partial x_1}(x) - \left[\frac{1}{C}x_1 + \frac{R}{L}x_2 - \beta(x)\right]\frac{\partial H}{\partial x_2}(x) = -\frac{1}{L}x_2\beta(x) \tag{2.57}$$

The energy function admits already as minima states where $x_2 = 0$ and thus it is enough to shape only the part of the energy that depends on x_1. Therefore, it can be considered a function $H_a = H_a(x_1)$. In this case Eq.(2.57) reduces to:

$$\beta(x_1) = -\frac{\partial H_a}{\partial x_1}(x_1) \tag{2.58}$$

which, for any $H_a(x_1)$ defines a control law $u = \beta(x_1)$. To shape the closed loop energy function in order to have a strict minimum in the configuration$x = (x_1^*, 0)^T$ it is enough to consider

$$H_a(x_1) = \frac{1}{2C_a}x_1^2 - (\frac{1}{C} + \frac{1}{C_a})x_1^*x_1 + k \tag{2.59}$$

The closed loop energy is the given by

$$H_d(x) = H(x) + H_a(x) = \frac{1}{2}(\frac{1}{C} + \frac{1}{C_a})(x_1 - x_1^*)^2 + \frac{1}{2L}x_2^2 + k \tag{2.60}$$

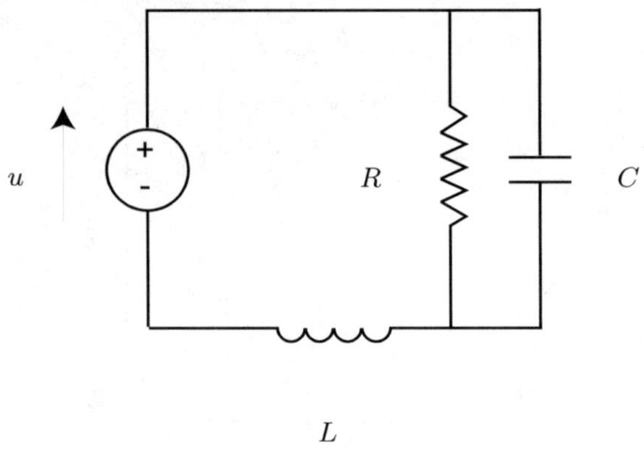

Fig. 2.3. Parallel RLC circuit

$H_d(x)$ has a minimum in the configuration $(x_1^*, 0)$ for all gains $C_a > -C$. Thus the control law:

$$u = -\frac{\partial H_a}{\partial x} = -\frac{x_1}{C_a} + (\frac{1}{C} + \frac{1}{C_a})x_1^* \tag{2.61}$$

stabilizes the configuration x^* and the Lyapunov function is given by the difference between the energy stored into the system and the energy supplied by the controller.

The energy-balance PBC works fine for the electrical example in Example 2.26 since the equilibrium configuration is characterized by no dissipation of energy and, therefore, the system can be stabilized by extracting a finite amount of power. On the other hand it is possible that some electrical circuits are not stabilizable with energy-balance PBC, as illustrated in the following example.

Example 2.27. Consider the parallel RLC circuit shown in Fig. 2.3. The energy variables are the charge q in the capacitor and the flux ϕ in the inductance; thus the state of the system can be chosen as $x = (q, \phi)^T = (x_1, x_2)^T$. The total energy function of the system is

$$H(x) = \frac{1}{2C}x_1^2 + \frac{1}{2L}x_2^2 \tag{2.62}$$

and the port-Hamiltonian model of the system is:

$$\begin{cases} \dot{x} = \begin{pmatrix} \dot{x}_1 \\ \dot{x}_2 \end{pmatrix} = \left[\begin{pmatrix} 0 & 1 \\ -1 & 0 \end{pmatrix} - \begin{pmatrix} R & 0 \\ 0 & 0 \end{pmatrix}\right] \begin{pmatrix} \frac{\partial H}{\partial x_1} \\ \frac{\partial H}{\partial x_2} \end{pmatrix} + \begin{pmatrix} 0 \\ 1 \end{pmatrix} u \\ y = (0\ 1) \begin{pmatrix} \frac{\partial H}{\partial x_1} \\ \frac{\partial H}{\partial x_2} \end{pmatrix} \end{cases} \tag{2.63}$$

The system satisfies Eq.(2.38) with $d(t) = R\int_0^t [(1/C)x_2^2(\tau)]^2 d\tau$, i.e. the energy dissipated by the resistor. It is clear from the constitutive equations that the admissible equilibrium configuration of the system are of the form $x^* = (Cu^*, (L/R)u^*)^T$. Notice that in correspondence of the equilibrium configuration there is dissipation and therefore infinite power should be extracted to stabilize the system. For this class of systems the energy-balance PBC is not working.

The energy-balance PBC can also be formulated for generic passive systems but the main drawback of this control technique, the dissipation obstacle, is present also in this more general framework; for further details see [227].

We have seen in Sec. 2.2.3 that the control by damping injection has a very clear physical interpretation since its action is physically equivalent to the presence of a dissipative element which, for example, in the mechanical domain can be modeled as a damper. The energy-based PBC changes, through a state feedback, the energy of the plant by giving it the desired shape. It is possible to give a physical interpretation to the energy-balance PBC control and to relate the dissipation obstacle to the physical properties of the plant. To this aim, it is necessary to give a more geometric interpretation of the control strategy and this can be done by using Willems' *control as interconnection.*

2.3.2 The Control as Interconnection Paradigm

In control theory, controllers are very often seen as *signal processors.* The controller receives from a set of sensors a set of information relative to the plant, processes them obeying to a certain control law and transmits them to a set of actuators that drive the plant. The energy-balance PBC described in Sec. 2.3.1 fits very well in this description: the controller receives the state x and processes it to obtain a control law $u = \beta(x) + v$ to stabilize the system at the desired admissible equilibrium configuration.

A novel way of looking at control and control problems is the so called *control as interconnection* which is based on the behavioral approach of modeling systems and that has been introduced in [330, 243]. In the following this control paradigm will be described for dynamical systems, which is the framework needed for the control of port-Hamiltonian systems, but it can be generalized for generic mathematical models; for further details see [330].

Definition 2.28 (Interconnection of dynamical systems). *Consider two continuous time dynamical systems* $\Sigma_1 = (\mathbb{R}, \mathcal{W}, \mathcal{B}_1)$ *and* $\Sigma_2 = (\mathbb{R}, \mathcal{W}, \mathcal{B}_2)$ *with the same signal space* $\mathcal{W}$*. The* interconnection of Σ_1 and Σ_2 *is denoted by* $\Sigma_1 \wedge \Sigma_2$ *and is a dynamical system defined as* $\Sigma_1 \wedge \Sigma_2 = (\mathbb{R}, \mathcal{W}, \mathcal{B}_1 \cap \mathcal{B}_2)$

Thus, given two dynamical systems $\Sigma_1 = (\mathbb{R}, \mathcal{W}, \mathcal{B}_1)$ and $\Sigma_2 = (\mathbb{R}, \mathcal{W}, \mathcal{B}_2)$ with the same signal space $\mathcal{W}$ the behavior of their interconnection consists of those trajectories $w : \mathbb{R} \to \mathcal{W}$ that are allowed both for Σ_1 (i.e. $w \in \mathcal{B}_1$) and for Σ_2 (i.e. $w \in \mathcal{B}_2$). Consequently, the trajectories of the interconnected systems must be acceptable for both systems and, thus, the interconnection

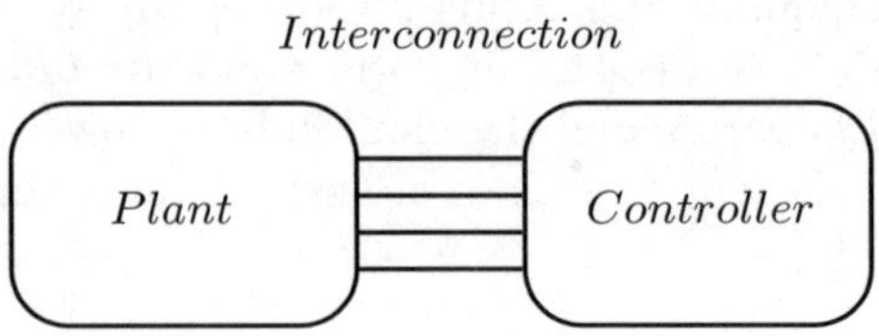

Fig. 2.4. Control as interconnection

limits the behaviors of both the interconnected systems to those trajectories that are compatible for both systems.

Consider a dynamical system $\Sigma_p = (\mathbb{R}, \mathcal{W}, \mathcal{B}_p)$ which represents the plant. Consider the set of all dynamical systems with $\mathbb{R}$ as time axis and with signal space $\mathcal{W}$; this set is called the *family of admissible controllers* and is denoted by $\mathfrak{C}$. An element $\Sigma_c \in \mathfrak{C}$ is called *admissible controller*. The interconnected system $\Sigma_p \wedge \Sigma_c$ is called the *controlled system*. Thus, give a plant Σ_p, the problem of control can thus be articulated in three steps:

1. Describe the set of admissible controllers $\mathfrak{C}$
2. Describe the properties that the controlled systems should have
3. Find an admissible controller $\Sigma_c \in \mathfrak{C}$ such that the system $\Sigma_p \wedge \Sigma_c$ has the desired properties.

Within the control as interconnection approach, controllers are no more signal processors but dynamical systems that are interconnected to the plant, as shown in Fig. 2.4. The behavior of the controlled system (i.e. of the interconnection between the plant and the controller) has to be obey to the laws imposed both by the plant and by the controller. The controller has to be designed in such a way that the undesired behaviors of the plant are no more allowed for the interconnected system, namely that the behavior of the plant when the controller is connected is exactly the desired one.

Sometimes it is more appropriate to see controllers as dynamical systems interconnected to the plant rather than simply signal processor. In this way the structural properties of the systems (geometric structures, invariants, etc.) can be fruitfully exploited to constrain the behavior of the plant to a desired subset and, concerning the control of port-Hamiltonian systems, it is possible to give a clear physical interpretation to the controller and to the dissipation obstacle.

Port-Hamiltonian systems interact with the external world by means of a power port and thus, the interconnection between port-Hamiltonian systems takes place through the respective power ports. Obviously, the energy exchange is fundamental in the design of passivity based controllers and, therefore, it is useful to consider a particular class of interconnections, namely the power preserving interconnections that allow, as shown in Sec. 1.4, to establish an energy transfer between the power ports interconnected. The situation is illustrated in Fig. 2.5.

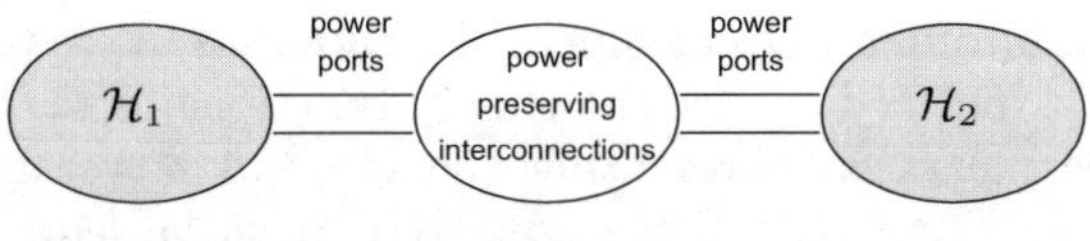

Fig. 2.5. Power preserving interconnection of port-Hamiltonian systems

The most general way to represent a set of power preserving interconnections is a Dirac structure, as shown in Sec. 1.4. On the other hand, for the control purposes that we are presenting, a power preserving interconnection can be represented with a skew-symmetric matrix $J_{int}(x_1, x_2)$, where x_1 and x_2 represent the states of the system 1 and of the system 2 respectively, which relates the inputs and the outputs of the interconnected systems:

$$\begin{pmatrix} u_1 \\ u_2 \end{pmatrix} = J_{int}(x_1, x_2) \begin{pmatrix} y_1 \\ y_2 \end{pmatrix} \tag{2.64}$$

It is immediate to prove that J_{int} represents a power preserving interconnection. In fact, recalling that in coordinates $\langle u, y \rangle = u^T y$:

$$u_1^T y_1 + u_2^T y_2 = \left[J_{int} \begin{pmatrix} y_1 \\ y_2 \end{pmatrix} \right]^T \begin{pmatrix} y_1 \\ y_2 \end{pmatrix} = - \begin{pmatrix} y_1^T & y_2^T \end{pmatrix} J_{int}(x_1, x_2) \begin{pmatrix} y_1 \\ y_2 \end{pmatrix} = 0 \tag{2.65}$$

which expresses the fact that there is a transfer of energy through the interconnection, namely, that the energy extracted from one system is supplied to the other without any loss or production of extra energy. The classic negative feedback interconnection

$$\begin{cases} u_1 = -y_2 \\ u_2 = y_1 \end{cases}$$

is an example of a power preserving interconnection where:

$$J_{int} = \begin{pmatrix} 0 & -I \\ I & 0 \end{pmatrix}$$

Let

$$\Sigma_1 = \begin{cases} \dot{x}_1 = (J_1(x_1) - R_1(x_1)) \dfrac{\partial H_1}{\partial x_1} + g(x_1) u_1 \\[2ex] y_1 = g_1^T(x_1) \dfrac{\partial H_1}{\partial x_1} \end{cases}$$
$$\Sigma_2 = \begin{cases} \dot{x}_2 = (J_2(x_2) - R_2(x_2)) \dfrac{\partial H_2}{\partial x_2} + g_2(x_2) u_2 \\[2ex] y_2 = g_2^T(x_2) \dfrac{\partial H_2}{\partial x_2} \end{cases} \tag{2.66}$$

be two port-Hamiltonian systems. $x_1 \in \mathcal{X}_1$ and $x_2 \in \mathcal{X}_2$ represent the state variables, $H_1(x_1)$ and $H_2(x_2)$ are the energy functions, $J_1(x_1)$ and $J_2(x_2)$ are skew symmetric matrices representing the internal power preserving interconnections, $R_1(x_1)$ and $R_2(x_2)$ are symmetric positive semidefinite matrices representing the dissipation of the system. Furthermore $(u_1, y_1) \in \mathcal{U}_1 \times \mathcal{Y}_1$ and $(u_2, y_2) \in \mathcal{U}_2 \times \mathcal{Y}_2$ are pair of power conjugated variables which describe the power ports by means of which each system can interact with the rest of the world. Let $J_{int}(x_1, x_2)$ be a skew symmetric matrix and consider the following power preserving interconnection:

$$\begin{pmatrix} u_1 \\ u_2 \end{pmatrix} = J_{int}(x_1, x_2) \begin{pmatrix} y_1 \\ y_2 \end{pmatrix} + \begin{pmatrix} v_1 \\ v_2 \end{pmatrix} \tag{2.67}$$

between the power ports (u_1, y_1) and (u_2, y_2).

Proposition 2.29. *The power preserving interconnection between the two port-Hamiltonian systems Σ_1 and Σ_2 yields another port-Hamiltonian system with state space given by the product space $\mathcal{X}_1 \times \mathcal{X}_2$ and with Hamiltonian function $H_1(x_1) + H_2(x_2)$.*

Proof. The matrix J_{int} can be partitioned in the following way:

$$J_{int}(x_1, x_2) = \begin{pmatrix} J_{11}(x_1, x_2) & J_{12}(x_1, x_2) \\ J_{21}(x_1, x_2) & J_{22}(x_1, x_2) \end{pmatrix} \tag{2.68}$$

where J_{11}, J_{12}, J_{21} and J_{22} are matrices of proper dimensions. From the skew symmetry of J_{int} it follows that the J_{11} and J_{22} are skew symmetric and that $J_{12}^T = -J_{21}$. By straightforward calculations it follows that the interconnected system takes the form:

$$\begin{cases} \begin{pmatrix} \dot{x}_1 \\ \dot{x}_2 \end{pmatrix} = \\ = [\begin{pmatrix} J_1(x_1) + g_1(x_1)J_{11}(x_1, x_2)g_1^T(x_1) & g_1(x_1)J_{12}(x_1, x_2)g_2^T \\ g_2(x_2)J_{21}(x_1, x_2)g_1^T(x_1) & J_2(x_2) + g_2(x_2)J_{22}(x_1, x_2)g_2^T(x_2) \end{pmatrix} - \\ \begin{pmatrix} R_1(x_1) & 0 \\ 0 & R_2(x_2) \end{pmatrix}] \begin{pmatrix} \frac{\partial H_1}{\partial x_1} \\ \frac{\partial H_2}{\partial x_2} \end{pmatrix} + \begin{pmatrix} g_1(x_1) & 0 \\ 0 & g_2(x_2) \end{pmatrix} \begin{pmatrix} v_1 \\ v_2 \end{pmatrix} \\ \begin{pmatrix} y_1 \\ y_2 \end{pmatrix} = \begin{pmatrix} g_1^T(x_1) & 0 \\ 0 & g_2^T(x_2) \end{pmatrix} \begin{pmatrix} \frac{\partial H_1}{\partial x_1} \\ \frac{\partial H_2}{\partial x_2} \end{pmatrix} \end{cases} \tag{2.69}$$

Since J_{11} and J_{22} are skew symmetric the terms $g_1(x_1)J_{11}(x_1, x_2)g_1(x_1)$ and $g_2(x_2)J_{22}(x_1, x_2)g_2(x_2)$ are also skew symmetric. Thus Eq.(2.69) represents a port-Hamiltonian system with state space $\mathcal{X} = \mathcal{X}_1 \times \mathcal{X}_2$ and with total energy function $H(x_1, x_2) = H_1(x_1) + H_2(x_2)$.

2.3.3 Energy Shaping as Control by Interconnection

In this section it is shown how to shape the energy of a plant, modeled as a port-Hamiltonian system, by interconnecting to it in a power preserving way, another port-Hamiltonian system, properly designed, which plays the role of the controller.
Consider a port-Hamiltonian system with dissipation:

$$\Sigma_P = \begin{cases} \dot{x} = (J(x) - R(x))\dfrac{\partial H}{\partial x} + g(x)u \\ \\ y = g^T(x)\dfrac{\partial H}{\partial x} \end{cases} \quad x \in \mathcal{X} \quad \dim \mathcal{X} = n \tag{2.70}$$

Definition 2.30 (Casimir function). *A Casimir function for a port-Hamiltonian system with dissipation in the form of Eq.(2.70) is a function $C : \mathcal{X} \to \mathbb{R}$, where $\mathcal{X}$ is the state space of the port-Hamiltonian system, that satisfies the following relation:*

$$\frac{\partial^T C}{\partial x}[J(x) - R(x)] = 0 \quad x \in \mathcal{X} \tag{2.71}$$

Casimir functions are constant along the trajectories of the unforced port-Hamiltonian system. In fact, letting $u = 0$ in Eq.(2.70), the following relation holds:

$$\dot{C}(x) = \frac{\partial^T C}{\partial x}[J(x) - R(x)]\frac{\partial H}{\partial x} = 0 \tag{2.72}$$

Thus Casimir functions are conserved quantities for the port-Hamiltonian systems. Nevertheless they are a very particular kind of conserved quantities since their variation along the trajectories of the unforced system is 0 *independently of* the Hamiltonian of the system. Thus, Casimir functions are conserved quantities that are determined *only* by the geometry of the system, namely by the interconnection structure ($J(x)$) and by the dissipation structure ($R(x)$).

Consider a port-Hamiltonian system with dissipation

$$\begin{cases} \dot{x}_c = [J_c(x_c) - R_c(x_c)]\,\frac{\partial H_c}{\partial x_c} + g_c(x_c)u_c \\ \\ y_c = g_c^T(x_c)\frac{\partial H_c}{\partial x_c} \end{cases} \quad x_c \in \mathcal{X}_c \quad \dim \mathcal{X}_c = n_c \tag{2.73}$$

regarded as a controller to interconnect to the plant described in Eq.(2.70) via the standard feedback interconnection:

$$\begin{pmatrix} u \\ u_c \end{pmatrix} = \underbrace{\begin{pmatrix} 0 & -I \\ I & 0 \end{pmatrix}}_{J_{int}} \begin{pmatrix} y \\ y_c \end{pmatrix} + \begin{pmatrix} v \\ v_c \end{pmatrix} \tag{2.74}$$

In Sec. 2.3.2 it has been shown that the the standard feedback interconnection is a power preserving interconnection and, furthermore, it has been proven that any power preserving interconnection between port-Hamiltonian systems with dissipation yields to another port-Hamiltonian system with dissipation. Therefore the closed loop port-Hamiltonian system deriving by the standard feedback interconnection between the plant and the controller is given by:

$$\begin{cases} \begin{pmatrix} \dot{x} \\ \dot{x}_c \end{pmatrix} = \left[\begin{pmatrix} J(x) & -g(x)g_c^T(x_c) \\ g_c(x_c)g^T(x) & J_c(x_c) \end{pmatrix} - \begin{pmatrix} R(x) & 0 \\ 0 & R_c(x_c) \end{pmatrix} \right] \begin{pmatrix} \frac{\partial H}{\partial x} \\ \frac{\partial H_c}{\partial x_c} \end{pmatrix} + \\ + \begin{pmatrix} g(x) & 0 \\ 0 & g_c(x_c) \end{pmatrix} \begin{pmatrix} v_1 \\ v_2 \end{pmatrix} \\ \begin{pmatrix} y \\ y_c \end{pmatrix} = \begin{pmatrix} g^T(x) & 0 \\ 0 & g_c^T(x_c) \end{pmatrix} \begin{pmatrix} \frac{\partial H}{\partial x} \\ \frac{\partial H_c}{\partial x_c} \end{pmatrix} \end{cases} \tag{2.75}$$

The state space of the closed loop system is $\mathcal{X} \times \mathcal{X}_c$ and $\dim(\mathcal{X} + \mathcal{X}_c) = n + n_c$ and the total energy is $H(x) + H_c(x_c)$

The Hamiltonian $H(x)$ of the plant is given while the Hamiltonian $H_c(x_c)$ of the plant can be freely assigned for control purposes. However, it is not very clear how a particular choice of the controller energy function can shape the total energy in such a way that the plant in stabilized in a certain configuration. The main idea of the energy shaping as a control by interconnection is to relate the state of the controller and the state of the plant through Casimir functions of the closed loop port-Hamiltonian system. In particular, Casimir functions of the following form are considered:

$$C_i(x, x_c) = x_{ci} - F_i(x) \qquad i = 1, \ldots, n_c \tag{2.76}$$

where $F_i : \mathcal{X} \to \mathbb{R}$ and x_{ci} denotes the ith component of x_c. Thus, each Casimir function relates a component of x_c to x.

By the definition of Casimir function and of the closed loop system, it follows that, for each function $C_i(x, x_c)$, the following partial differential equations have to hold:

$$\begin{aligned} &\frac{\partial^T F_i}{\partial x}(x)[J(x) - R(x)] - g_c^i(x_c)g^T(x) = 0 \\ &\frac{\partial^T F_i}{\partial x} g(x)g_c^T(x_c) + J_c^i(x_c) - R_c^i(x_c) = 0 \end{aligned} \qquad i = 1, \ldots, n_c \tag{2.77}$$

where g_c^i, J_c^i and R_c^i denote the ith row of g_c, J_c and R_c respectively. In order to relate x_c to x, n_c Casimir functions have to be found and thus the following set of partial differential equations has to be found:

$$\frac{\partial^T F}{\partial x}(x)[J(x) - R(x)] - g_c(x_c)g^T(x) = 0 \tag{2.78a}$$

$$\frac{\partial^T F}{\partial x} g(x) g_c^T(x_c) + J_c(x_c) - R_c(x_c) = 0 \tag{2.78b}$$

where $F = (F_1, \ldots, F_{n_c})^T$.
The post-multiplication of Eq.(2.78a) by $\frac{\partial F}{\partial x}$ and using Eq.(2.78b) yields:

$$\frac{\partial^T F}{\partial x} [J(x) - R(x)] \frac{\partial F}{\partial x} = J_c(x_c) + R_c(x_c) \tag{2.79}$$

In general, given two skew symmetric matrices J_1 and J_2, and two symmetric matrices R_1 and R_2, $J_1 + R_1 = J_2 + R_2$ implies $J_1 = J_2$ and $R_1 = R_2$ [117]. Therefore Eq.(2.79) can be rewritten as:

$$\frac{\partial^T F}{\partial x} J(x) \frac{\partial F}{\partial x} = J_c(x_c) \tag{2.80}$$

$$-\frac{\partial^T F}{\partial x} R(x) \frac{\partial F}{\partial x} = R_c(x_c) \tag{2.81}$$

Since $R_c(x_c)$ and $R(x)$ are symmetric positive semidefinite, Eq.(2.81) can be rewritten as:

$$\frac{\partial^T F}{\partial x} R(x) \frac{\partial F}{\partial x} = 0 \tag{2.82}$$

$$R_c(x_c) = 0 \tag{2.83}$$

Since $R(x)$ is symmetric positive semidefinite, Eq.(2.82) is equivalent to

$$R(x) \frac{\partial F}{\partial x} = 0 \tag{2.84}$$

Finally, using Eq.(2.84), it is possible to rewrite Eq.(2.78a) as:

$$\frac{\partial^T F}{\partial x}(x) J(x) = g_c(x_c) g^T(x) \tag{2.85}$$

Therefore the following result has been just proven [227]:

Proposition 2.31. *The functions $C_i(x, x_c) = x_{ci} - F_i(x)$, $i = 1, \ldots, n_c$ are Casimir functions for the closed loop port-Hamiltonian system in Eq.(2.75) if and only if $F(x) = (F_1(x), \ldots, F_{n_c}(x)^T)$ satisfies the following partial differential equations:*

$$\begin{aligned}
&\frac{\partial^T F}{\partial x} J(x) \frac{\partial F}{\partial x} = J_c(x_c) \\
&R(x) \frac{\partial F}{\partial x} = 0 \\
&R_c(x_c) = 0 \\
&\frac{\partial^T F}{\partial x}(x) J(x) = g_c(x_c) g^T(x)
\end{aligned} \tag{2.86}$$

The Casimir functions are conserved along the trajectories of the unforced closed loop port-Hamiltonian system which is therefore constrained to evolve on the set:

$$L_C = \{(x, x_c) \mid x_{ci} = F_i(x) + c_i,\ i = 1, \dots n_c\} \tag{2.87}$$

where c_i is the constant value that each Casimir function C_i assumes.

The dynamics of the plant in the unforced closed loop controlled system represented in Eq.(2.75) is give by:

$$\dot{x} = [J(x) - R(x)]\frac{\partial H}{\partial x}(x) - g(x)g_c^T(x_c)\frac{\partial H_c}{\partial x_c}(x_c) \tag{2.88}$$

which, using the second and the fourth relation of Eq.(2.86), can be rewritten as

$$\dot{x} = [J(x) - R(x)](\frac{\partial H}{\partial x}(x) + \frac{\partial F}{\partial x}(x)\frac{\partial H_c}{\partial x_c}(x_c)) \tag{2.89}$$

But:

$$x_c = F(x) + c \tag{2.90}$$

where $c = (c_1, \dots, c_{n_c})^T$. Using the chain rule for differentiation:

$$\frac{\partial H_c(F(x) + c)}{\partial x} = \frac{\partial F}{\partial x}(x)\frac{\partial H_c}{\partial x_c}(F(x) + c) \tag{2.91}$$

Thus the dynamics of the controlled plant is given by:

$$\dot{x} = [J(x) - R(x)]\frac{\partial H_s}{\partial x}(x) \tag{2.92}$$

where

$$H_s(x) = H(x) + H_c(F(x) + c) \tag{2.93}$$

Therefore, the feedback interconnection of the plant in Eq.(2.70) with the controller in Eq.(2.73) such that there exist n_c Casimir functions of the form of Eq.(2.76) is again a port-Hamiltonian system with dissipation with the same interconnection and dissipation structure but with a *shaped* Hamiltonian function given by Eq.(2.93).

The energy of the controlled plant can be shaped by a proper selection of the Hamiltonian function of the controller that has to be chosen in such a way that $H_s(x)$ has a strict minimum in the desired configuration. It is straightforward to give an energy-balance PBC interpretation to the energy shaping control algorithm just illustrated. Since $R_c(x_c)$ has to be zero by Eq.(2.83)

$$\frac{dH_c}{dt} = u_c^T y_c \tag{2.94}$$

and thus, since $u = -y_c$ and $y = u_c$

$$\frac{dH_s}{dt} = \frac{dH}{dt} + \frac{dH_c}{dt} = \frac{dH}{dt} - u^T y \tag{2.95}$$

which yields:

$$H_s(x(t)) = H(x(t)) - \int_0^t u^T(\tau)y(\tau)d\tau \tag{2.96}$$

namely, the shaped energy is the difference between the energy stored in the plant and the energy supplied by the controller.

Facing the problem of energy shaping for port-Hamiltonian systems with dissipation as a control by interconnection problem has several advantages. First, it is possible to give a nice physical interpretation to the controller. In fact it can be thought as a physical extension of the plant designed in such a way that the behavior of the plant is constrained to the desired one. Second, the geometric structures characterizing port-Hamiltonian systems can be explicitly exploited for the design of the controller. In other words, all the energetic properties characterizing physical plants are explicitly used for building the control law and, therefore, physics plays a very active role in the determination of control which is no more reduced to a mere mathematical problem. Third, the drawbacks and the major advantages of the control algorithm admit a nice physical interpretation which cannot always be made clear in a state feedback framework.

The energy-shaping implemented by means of Casimir functions is equivalent to the control algorithm illustrated in Sec. 2.3.1 and therefore the dissipation obstacle is still the main drawback of the control strategy. On the other hand, it is now possible to characterize the admissible dissipation; in fact, since Eq.(2.84) must hold, the following relation holds as well:

$$R(x)\frac{\partial H_c(F)}{\partial x}(x) = 0 \tag{2.97}$$

Thus, a port-Hamiltonian system is stabilizable by means of energy-balancing PBC if and only if the dissipation structure of the plant and the Hamiltonian of the controller satisfies Eq.(2.97). Loosely speaking, energy balance PBC is possible only when H_c does not depend on the coordinates where there is dissipation; in other words,the coordinates on which there is natural dissipation need not to be shaped. In fact it is possible to stabilize mechanical systems in a determined configuration since the coordinate that need to be shaped is the position while dissipation in mechanical systems is present on momenta.

2.4 Interconnection and Damping Assignment Passivity Based Control

In Sec. 2.3.3 it has been shown that to shape the energy of the plant it is necessary that the closed loop system admits Casimir functions. While this is not a problem for the regulation of mechanical systems, it can become a concern when dealing with the control of electrical machines and, consequently, when explicitly considering the electric actuation system of a robot. The existence of Casimir functions imposes several conditions which are reported in Eq.(2.86). The main limitation imposed by these condition is on the admissible dissipation of the plant.

The dissipation obstacle expresses the fact that a port-Hamiltonian system cannot be stabilized in an admissible equilibrium configuration characterized by an infinite amount of dissipation by supplying a finite amount of energy. Thus, to stabilize a system in equilibrium points with infinite dissipation, controllers being able to supply an infinite amount of energy have to be considered. The controller can be describe by a port-Hamiltonian system characterized by a lower unbounded energy function as:

$$\Sigma_c = \begin{cases} \dot{x}_c = u_c \\ y_c = \frac{\partial H_c}{\partial x_c}(x_c) \end{cases} \tag{2.98}$$

with energy function

$$H_c(x_c) = -x_c \tag{2.99}$$

Furthermore, it has been shown in Sec. 2.3.3 that the standard feedback interconnection implies very strict conditions, related to the existence of Casimir functions, expressed by Eq.(2.86). It is possible to get rid of these restricting conditions by embedding the plant state information into the interconnection. Thus the following power preserving interconnection can be considered:

$$\begin{pmatrix} u(t) \\ u_c(t) \end{pmatrix} = \underbrace{\begin{pmatrix} 0 & -\beta(x) \\ \beta^T(x) & 0 \end{pmatrix}}_{J_{int}(x)} \begin{pmatrix} y(t) \\ y_c(t) \end{pmatrix} \tag{2.100}$$

The system resulting from the coupling of the systems in Eq.(2.70) and in Eq.(2.98) through the interconnection represented by Eq.(2.100) is clearly a port-Hamiltonian system with dissipation, because of Proposition 2.29, and has the form:

$$\begin{pmatrix} \dot{x} \\ \dot{x}_c \end{pmatrix} = \left[\begin{pmatrix} J(x) & -g(x)\beta(x) \\ \beta^T(x)g^T(x) & 0 \end{pmatrix} - \begin{pmatrix} R(x) & 0 \\ 0 & 0 \end{pmatrix} \right] \begin{pmatrix} \frac{\partial H}{\partial x}(x) \\ \frac{\partial H_c}{\partial x_c}(x_c) \end{pmatrix} \tag{2.101}$$

It is possible to shape the energy of the closed loop system *without* the generation on Casimir functions and thus without the restrictions imposed by the necessary conditions reported in Eq.(2.86), in particular without the dissipation obstacle. In fact, if it is possible to solve the following partial differential equation:

$$[J(x) - R(x)]\frac{\partial H_a}{\partial x}(x) = g(x)\beta(x) \tag{2.102}$$

for some $\beta(x)$ the plant dynamics will be:

$$\dot{x} = [J(x) - R(x)]\frac{\partial H_d}{\partial x}(x) \tag{2.103}$$

where $H_d(x) = H(x) + H_a(x)$. If it is possible to choose $H_a(x)$ such that $H_d(x)$ as a minimum in the desired configuration, then the interconnection of Eq.(2.70) with Eq.(2.98) through Eq.(2.100) will be asymptotically stabilized in the desired configuration. Notice that there are not conditions to fulfill and, in particular, no dissipation obstacle. Thus, this new control scheme is applicable also for the stabilization of infinite dissipation systems.

Notice that controlling the plant through the interconnection of the controller in Eq.(2.98) by means of Eq.(2.100) is equivalent to control it by means of the static feedback $u = \beta(x)$. In fact, from Eq.(2.101) it follows that:

$$\dot{x} = [J(x) - R(x)]\frac{\partial H}{\partial x}(x) - g(x)\beta(x)\frac{\partial H_c}{\partial x_c} \tag{2.104}$$

which, since $\frac{\partial H_c}{\partial x_c} = -1$, is equivalent to:

$$\dot{x} = [J(x) - R(x)]\frac{\partial H}{\partial x}(x) + g(x)\beta(x) \tag{2.105}$$

Example 2.32. Consider again the parallel RLC circuit represented in Fig. 2.3. The system can be modeled as a port-Hamiltonian system with dissipation, as reported in Eq.(2.63), with energy function reported in Eq.(2.62). The admissible equilibrium configurations of the system are of the form $x^* = (Cu^*, (L/R)u^*)^T$ and the system is characterized by an infinite dissipation as shown in Example 2.27. The partial differential equation reported in Eq.(2.102) applied to this system becomes:

$$-\frac{1}{R}\frac{\partial H_a}{\partial x_1}(x) + \frac{\partial H_a}{\partial x_2}(x) = 0 \tag{2.106a}$$

$$-\frac{\partial H_a}{\partial x_1}(x) = \beta(x) \tag{2.106b}$$

A solution of Eq.(2.106a) is:

$$H_a(x) = \Phi(Rx_1 + x_2) \tag{2.107}$$

where $\Phi : \mathbb{R} \to \mathbb{R}$ is an arbitrary differentiable function. Then, once $H_a(x)$ has been determined, Eq.(2.106b) defines the interconnection. The system has to be stabilized in an admissible equilibrium configuration and thus Φ has to be chosen in such a way that $H_d(x) = H(x) + H_a(x)$ has a minimum in the configuration $x^* = (Cu^*, (L/R)u^*)^T$. A possible solution is:

$$\Phi(Rx_1 + x_2) = \frac{K_p}{2}[(Rx_1 + x_2) - (Rx_1^* + x_2^*)]^2 - Ru^*(Rx_1 + x_2) \tag{2.108}$$

which renders x^* a minimum of $H_d(x)$ for any

$$K_p > -\frac{1}{(L + CR)^2}$$

The energy function $H_d(x)$ is:

$$H_d(x) = (x - x^*)^T \begin{pmatrix} \frac{1}{C} + R^2 K_p & RK_p \\ RK_p & \frac{1}{L} + K_p \end{pmatrix} (x - x^*) \tag{2.109}$$

which has a global minimum in the desired configuration.

It has been shown that it is possible to shape the energy of the closed loop port-Hamiltonian system with dissipation *without* the generation of Casimir functions and, therefore, also for systems with infinite dissipation. On the other hand, the implementation of the control strategy relies on the solution of Eq.(2.102) and it is well known that it is not easy to solve, in general, partial differential equations. It is possible to simplify Eq.(2.102) using the extra degrees of freedom provided by the port-Hamiltonian structures of the plant and of the controller. In fact, for the stabilization of an admissible equilibrium configuration, it is only required to shape the energy of the closed loop port-Hamiltonian system in such a way that the desired configuration is a strict minimum for the new energy function. Using the techniques illustrated so far the closed loop dynamics of the plant is:

$$\dot{x} = [J(x) - R(x)]\frac{\partial H_d}{\partial x}(x) \tag{2.110}$$

where the interconnection and damping matrices are exactly those of the plant and $H_d(x)$ is the properly shaped energy function. From an energy shaping point of view nothing changes if the following closed loop dynamics is obtained:

$$\dot{x} = [J_d(x) - R_d(x)]\frac{\partial H_d}{\partial x} \tag{2.111}$$

where $J_d(x)$ and $R_d(x)$ are the desired interconnection and damping matrix and $H_d(x)$ is the shaped energy function. Thus it is possible to freely assign an interconnection matrix and a damping matrix for the closed loop system without affecting the stability properties of the desired configuration. This extra degrees of freedom lead to the so-called *Interconnection and Damping*

Assignment Passivity Based Controllers (IDA-PBC) [227, 229]. In this case Eq.(2.102) becomes:

$$[J(x) + J_a(x) - R(x) - R_a(x)]\frac{\partial H_a}{\partial x}(x) = -[J_a(x) - R_a(x)]\frac{\partial H}{\partial x}(x) + g(x)\beta(x) \tag{2.112}$$

where

$$J_a(x) = J_d(x) - J(x), \quad R_a(x) = R_d(x) - R(x)$$

Thus it is possible to choose a desired closed loop interconnection matrix and a desired closed loop damping matrix in order to simplify Eq.(2.112) and to get a solution for the controller design. The following proposition can be proven [229]:

Proposition 2.33. *Consider a port-Hamiltonian system with dissipation characterized by the matrices $J(x)$, $R(x)$, $g(x)$ and by the energy function $H(x)$. Let x^* be a configuration to be stabilized and assume that it is possible to find functions $\beta(x)$, $R_a(x)$ and $J_a(x)$ such that:*

$$J(x) + J_a(x) = -[J(x) + J_a(x)]^T$$

$$R(x) + R_a(x) = [R(x) + R_a(x)]^T \quad \textit{positive semidefinite}$$

and a vector function $K(x)$ such that:

$$[J(x) + J_a(x) - R(x) - R_a(x)]\frac{\partial K}{\partial x}(x) = -[J_a(x) - R_a(x)]\frac{\partial H}{\partial x}(x) + g(x)\beta(x) \tag{2.113}$$

and such that the following conditions are satisfied:

1. (Integrability)
$$\frac{\partial K}{\partial x}(x) = \frac{\partial^T K}{\partial x}(x)$$

2. (Equilibrium Assignment)
$$K(x^*) = -\frac{\partial H}{\partial x}(x^*)$$

3. (Lyapunov Stability)
$$\frac{\partial K}{\partial x}(x^*) > -\frac{\partial^2 H}{\partial x^2}(x^*)$$

Under these conditions, the static feedback $u = \beta(x)$ is a port-Hamiltonian system with dissipation and the dynamic of the plant is:

$$\dot{x} = [J_d(x) - R_d(x)]\frac{\partial H_d}{\partial x}$$

where $H_d(x) = H(x) + H_a(x)$ *and*

$$\frac{\partial H_a}{\partial x}(x) = K(x)$$

Furthermore, x^* *is a (locally) stable equilibrium of the closed loop.*

The IDA-PBC control technique generalizes the energy-balancing control strategy and allows to shape the energy for a much broader set of port-Hamiltonian systems with dissipation exploiting all the degrees of freedom given by the port-Hamiltonian formalism. This strategy is useful for regulating the behavior of non fully actuated robots, see for example [2].

2.5 A Variable Structure Approach to Energy-Based Control

A possible drawback of the energy shaping technique is that an exact knowledge of the system physical parameters is needed to correctly shape the energy function: this is not always true in practical applications. A consequence is that the energy function does not assume its minimum in the desired configuration and some regulation errors are introduced.

In this section, it is shown how to overcome this drawback by introducing a variable structure controller that properly shapes the total energy of the system while keeping on guaranteeing that the minimum of the total energy is in the desired configuration despite of possible parametric uncertainties. Furthermore, the passivity (and therefore the stable behavior) of the overall system will be preserved, [177]. We will illustrate the control strategy for the control of robot manipulators and we will provide simulation results on a 2-DOF planar manipulator, also taking into account a possible saturation of the actuators. With a very little effort, all the results obtained in this section can be generalized to the control of generic port-Hamiltonian systems, see for example [178, 256].

Consider again the n-DOF fully actuated mechanical system described in Example 2.25 and represented by Eq.(2.46).

Suppose that $q_d \in \mathcal{Q}$ is a desired configuration in the joint space. Following the steps reported in Example 2.25 and by introducing some further dissipation by damping injection, it is possible to design the following control action:

$$u = B^{-1}\left[\frac{\partial V}{\partial q} - \frac{\partial H_c}{\partial q} - K_D\frac{\partial H}{\partial p}\right] \tag{2.114}$$

where $H_c(\cdot)$ is the desired potential energy for the closed loop system. The state feedback law (2.114) shapes the total energy by compensating the effect of the potential $V(q)$ and by introducing a new potential $H_c(q)$. If H_c is characterized by a minimum in the desired configuration q_d, by introducing a

dissipative effect with the controller this new minimum is made asymptotically stable.
If we consider

$$H_c(q) = \frac{1}{2}(q - q_d)^T K_P (q - q_d) \tag{2.115}$$

with $K_P = K_P^T > 0$ the feedback law becomes

$$u(q,p) = B^{-1}\left[\frac{\partial V}{\partial q} - K_P(q - q_d) - K_D\dot{q}\right]$$

that is the well-known PD plus gravity compensation (PD + g(q)) controller, [12, 164]. Moreover, the closed-loop energy function becomes

$$H_{cl}(q,p) = \frac{1}{2}p^T M^{-1}(q)p + \frac{1}{2}(q - q_d)^T K_P (q - q_d) \tag{2.116}$$

The PD + g(q) controller can be interpreted as a set on n linear springs acting in the joint space with minimum in q_d. In order to extend this controller to take into account the saturation of each actuator, a non-linearity in the behavior of the springs has to be introduced.

It is well known that a spring is an element storing potential energy and its behavior can be described as shown in Fig. 2.6. The input u is the deformation

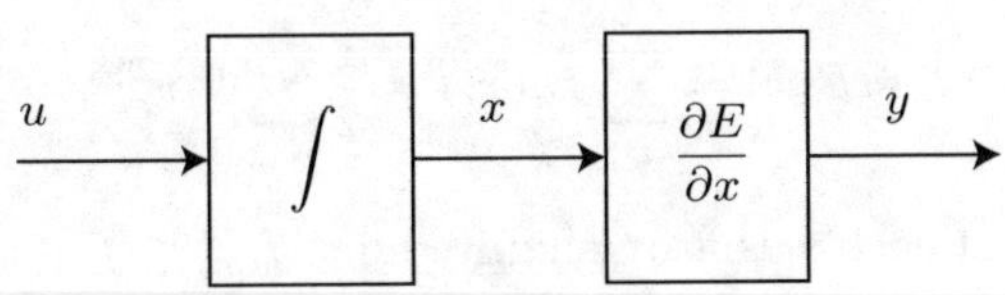

Fig. 2.6. Energy storing element behavior.

rate of the extreme of the spring, x is the state associated to the spring and $E(x)$ is a lower bounded function representing the stored energy. The output y is the force applied by the spring. The simplest springs are the linear ones, i.e. springs whose energy function is quadratic:

$$E(x) = \frac{1}{2}x^T K x \tag{2.117}$$

where $K = K^T > 0$ represents the stiffness. The force applied by the springs turns out to be:

$$f = \frac{\partial E}{\partial x} = Kx$$

In the case of mechanical systems (e.g. robots), each component of the force is applied to the plant by means of an actuator. Intuitively speaking, if the

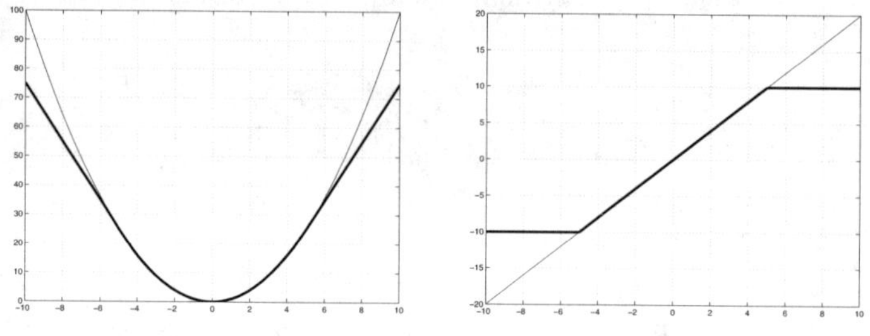

(a) Energy (b) Force

Fig. 2.7. Energy and force of non saturated (regular) and saturated (bold) spring.

amount of stored energy increases too much, then the force generated by the springs, that is the force that the actuators should apply, can be greater than the physical limits of the actuators themselves. If the robot is controlled by means of the PD + g(q) controller, this situation can happen e.g. if the initial error is sufficiently high.
Consider $K = \text{diag}(k_1, \ldots, k_n)$, that is the spring energy in (2.117) can be written as

$$E(x) = \sum_{i=1}^{n} E_i(x_i) = \frac{1}{2} \sum_{i=1}^{n} k_i x_i^2$$

Then, suppose that each actuator is limited, i.e. $f_{i,m} \leq f_i \leq f_{i,M}$, $i = 1, \ldots, n$. Consider $x_M = (x_{1,M}, \ldots, x_{n,M})$ and $x_m = (x_{1,m}, \ldots, x_{n,m})$ such that $f_{i,M} = k_i x_{i,M}$ and $f_{i,m} = k_i x_{i,m}$. The saturation of each actuator can be taken into account if the following energy function is introduced:

$$E_s(x) = E_{1,s} + \cdots + E_{n,s} \tag{2.118}$$

where

$$E_{i,s}(x_i) = \begin{cases} f_{i,m} \left[x_i - \frac{1}{2} x_{i,m} \right] & x_i < x_{i,m} \\ \frac{1}{2} k_i x_i^2, & x_{i,m} \leq x_i \leq x_{i,M} \\ f_{i,M} \left[x_i - \frac{1}{2} x_{i,M} \right] & x_i > x_{i,M} \end{cases} \tag{2.119}$$

Note that the passivity properties of the spring are preserved since the proposed energy function is $\mathcal{C}^1$ and bounded from below.

The energy function of a 1–dimensional spring and the relative force in function of the state are represented in Fig. 2.7, both for the non-saturated and saturated case. Clearly, the energy functions are different in the saturation zone: when the stored energy becomes infinite, the force generated by a non-saturated spring increases to infinity, while it is limited for the saturated case.

The saturation of each actuator can be taken into account in the (passive) control of a robot if, instead of the energy function (2.115), we consider it is assumed

$$H_c(q) = \sum_{i=1}^{n} E_{i,s}(q_i - q_{i,d}) \tag{2.120}$$

where $E_{i,s}(\cdot)$ is defined as in (2.119), $k_i > 0$ can be freely assigned, and $f_{i,m}$, $f_{i,M}$ depend on the characteristics of the i-th actuator. Since $H_c(\cdot)$ is characterized by a (global) minimum in q_d, the control action (2.114) still assures the (global) stability of this configuration.
From Eq.(2.116), it follows that

$$\frac{dH_{cl}}{dt} = -\frac{\partial^T H_{cl}}{\partial p}(D + K_D)\frac{\partial H_{cl}}{\partial p} < 0 \tag{2.121}$$

for $\frac{\partial H_{cl}}{\partial p} = \dot{q} \neq 0$. Since H_{cl} is bounded from below, we have that, for every $\epsilon > 0$, it is possible to find t_ϵ such that $\dot{q}(t) < \epsilon$ for every $t \geq t_\epsilon$. Moreover, the possible configurations in which the robot stops are clearly given by the solutions of:

$$\left.\frac{\partial H_{cl}}{\partial q}(q,p)\right|_{p=0} = 0 \tag{2.122}$$

or, in the case of perfect compensation of the potential $V(q)$, of:

$$\frac{\partial H_c}{\partial q}(q) = 0 \tag{2.123}$$

Therefore, if H_c is characterized by a global minimum in $q = q_d$, e.g. as in (2.115) or in (2.120), the robot reaches the desired configuration q_d. The key point is that a perfect compensation of the original potential energy of the robot has to be implemented. If this is not the case, then some regulation errors will be present.
Suppose that $\hat{V}(q)$ is an estimate of the potential term in $H(q,p)$. Then, (2.114) becomes

$$u = B^{-1}\left[\frac{\partial \hat{V}}{\partial q} - \frac{\partial H_c}{\partial q} - K_D\frac{\partial H}{\partial p}\right] \tag{2.124}$$

and H_{cl} is now given by

$$H_{cl}(q,p) = \frac{1}{2}p^T M^{-1}(q)p + H_c(q) - \Delta V(q) \tag{2.125}$$

where $\Delta V(q) = \hat{V}(q) - V(q)$. Since (2.121) holds, the final configurations the robot can assume are still solutions of (2.122), or, equivalently, of

$$\frac{\partial H_c}{\partial q} = \frac{\partial \Delta V}{\partial q} \tag{2.126}$$

Even if H_c is characterized by a (global) minimum in q_d, it is not sure that this configuration can be reached.

In order to make the control law (2.124) robust also in terms of *performances* with respect to unknown parameters, H_c, which is freely assignable, can be chosen with a variable structure. For example, assume

$$H_c(q) = \frac{1}{2}\sum_{i=1}^{n} k_i[q_i - q_{i,d} + \text{sign}(q_i - q_{i,d})\bar{q}_i]^2 \tag{2.127}$$

where $k_i > 0$ and $\bar{q}_i > 0$, with $i = 1, \ldots, n$. It is possible to prove that, if

$$\left|\frac{\partial \Delta V}{\partial q}\right| \leq M < \infty \tag{2.128}$$

and if $\bar{q}_i$, $i = 1, \ldots, n$, are *properly* chosen, then the control law (2.124) with H_a given by (2.127), can drive the system in $q = q_d$. The proof is immediate in the case that a perfect compensation of the potential $V(q)$ is possible, that is if $\Delta V(q) = 0$: in fact, in this situation, H_{cl} is characterized by a global minimum in $(q_d, 0)$.

Suppose, then, that $\Delta V(q) \neq 0$ and, in particular, that (2.128) holds and consider a generic initial condition (q_0, p_0). Define $\sigma := [\sigma_1, \ldots, \sigma_n]$, where

$$\sigma_i = \begin{cases} 1 & \text{if } q_{i,0} - q_{i,d} \geq 0 \\ -1 & \text{if } q_{i,0} - q_{i,d} < 0 \end{cases}$$

only depends on the initial condition. Then, assume that the control input u is given by (2.124), but with H_c given by:

$$H_c(q) = \frac{1}{2}\sum_{i=1}^{n} k_i(q_i - q_{i,d} + \sigma_i\bar{q}_i)^2 \tag{2.129}$$

If, with a proper choice of $\bar{q}$, this continuous control input can drive the robot in a final configuration q^* such that

$$\begin{cases} q_i^* - q_{i,d} < 0 \text{ if } q_{i,0} - q_{i,d} > 0 \;\; (\sigma_i = 1) \\ q_i^* - q_{i,d} > 0 \text{ if } q_{i,0} - q_{i,d} < 0 \;\; (\sigma_i = -1) \end{cases} \tag{2.130}$$

then an instant $\bar{t}$ such that $q(\bar{t}) - q_d = 0$ has to exists. Consequently, the variable structure controller resulting from (2.124) and (2.127) makes the configuration $q = q_d$ globally attractive and, clearly, globally stable.
The possible final configurations q^* are solution of (2.126), that is

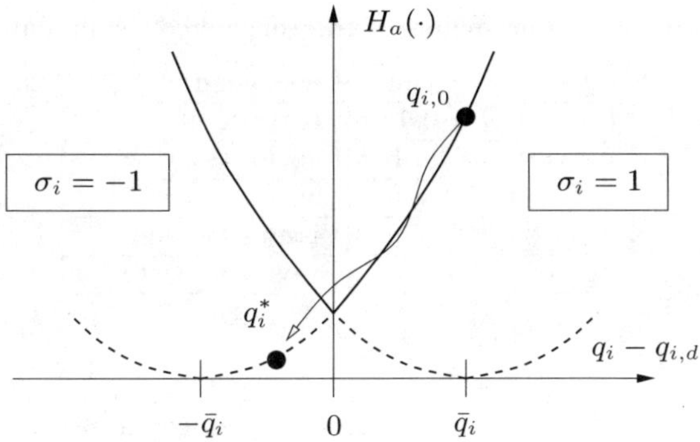

Fig. 2.8. Behavior of the proposed controller.

$$q_i^* - q_{i,d} + \sigma_i \bar{q}_i = \frac{1}{k_i} \frac{\partial \Delta V}{\partial q}(q^*) \tag{2.131}$$

Since the values $\bar{q}_i$, $i = 1, \ldots, n$ have to be chosen according to (2.130), it should be

$$\bar{q}_i > \frac{M}{k_i}, \qquad i = 1, \ldots, n \tag{2.132}$$

With this choice, the configuration $q = q_d$ is globally attractive and stable. In Fig. 2.8, the behavior of the proposed controller is illustrated: the initial error $q_{i,0} - q_{i,d}$ is greater than 0, but $\bar{q}_i$ is chosen in such a way that all the possible steady state configurations q^* satisfy $q_i^* - q_{i,d} < 0$ if H_c is given by (2.129). If the variable structure of the controller deriving from (2.127) is adopted, then the system is *constrained* in q_d.

The actuator saturation can be taken into account by introducing the saturated springs. If $E_{i,s}$ is the energy function of a saturated spring as reported in (2.119), suppose that

$$H_c(q) = \sum_{i=1}^{n} E_{i,s}[q_i - q_{i,d} + \text{sign}(q_i - q_{i,d})\bar{q}_i] \tag{2.133}$$

Then, the final configurations the robot can assume if u is given by (2.124) are the solutions of (2.126). If

$$|\max(f_{i,m}, f_{i,M})| > M \left(\geq \frac{\partial \Delta V}{\partial q} \right) \tag{2.134}$$

$i = 1, \ldots, n$, then in the steady state configuration none of the actuators is in saturation. A consequence is that, if $\bar{q}_i$, $i = 1, \ldots, n$, are chosen according to (2.132), then the controller is able to regulate the robot in q_d that will be an asymptotically stable configuration.

Table 2.1. Parameters of the considered manipulator.

$L_1 = L_2 = 1$ m	Links lengths
$L_{g1} = L_{g2} = 0.5$ m	Center of mass
$M_1 = M_2 = 20$ Kg	Links mass
$I_1 = I_2 = 5$ Kg m^2	Links inertia
$D_1 = D_2 = 0$ N m s	Viscous friction
$g = 9.81$ m / s^2	Gravity acceleration

In conclusion, even in presence of modeling uncertainties, a variable structure passive controller (2.124), with H_a given by (2.127) or (2.133), if the saturation of the actuators is taken into account, is able to drive the system in the desired configuration.

In order to test the control proposed algorithm, a 2-DOF planar manipulator has been considered, as shown in Fig. 2.9. The main parameters of the manipulator are reported in Tab. 2.1. Note that the manipulator is subject to gravity force active in the negative y direction. In the following, simulation results are reported in order to show the features of the proposed controller in comparison with the classical PD + gravity compensation regulator. As a reference case, in Fig. 2.10 a simulation with the PD + g(q) compensation controller is reported. A fixed set point $(x_d, y_d) = (1.75,\ 0.1)$ has been assigned as desired goal for the tip of the manipulator, corresponding to joint positions $q_1 = -25.57, q_2 = 57.57$ (deg). In this case, the dynamic parameters are supposed to be perfectly known. As expected, the errors nicely tend to zero, as shown in Fig. 2.10. In this case, the control parameters are $K_p = diag(6000, 6000)$, $K_d = diag(1100, 1100)$. The final errors are $e_x = -0.0001$, $e_y = 0.000064$ (m) corresponding to $e_{q_1} = 0.0047$, $e_{q_2} = -0.0141$ (deg). Results obtained with the proposed controller are reported in Fig. 2.11 and Fig. 2.12, with the same control parameters as in the previous case for the PD part,

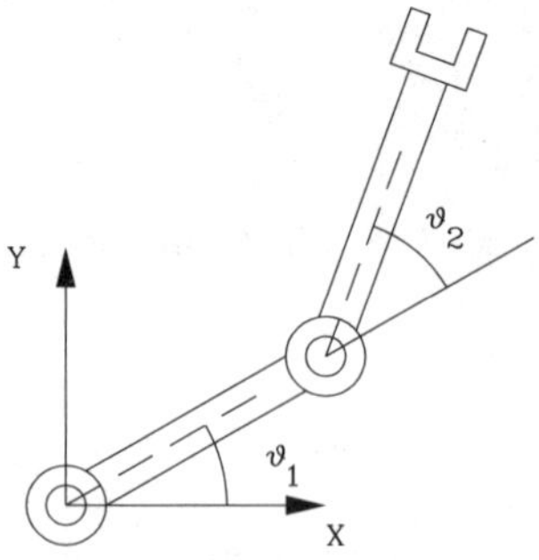

Fig. 2.9. A planar 2-DOF manipulator

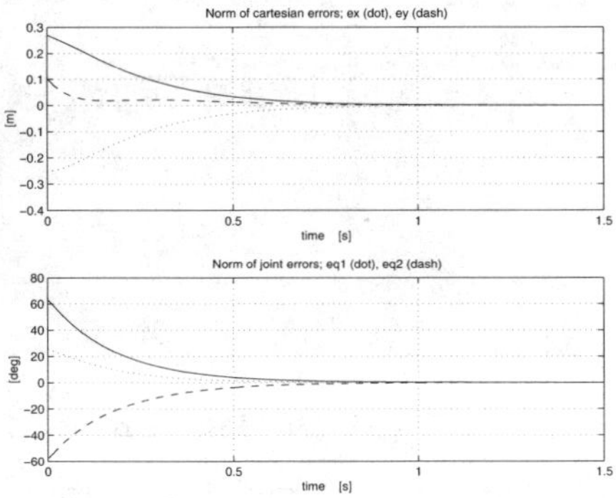

(a) Errors

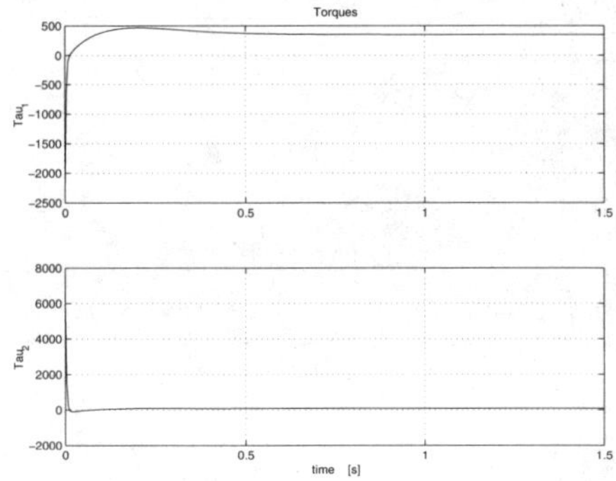

(b) Torques.

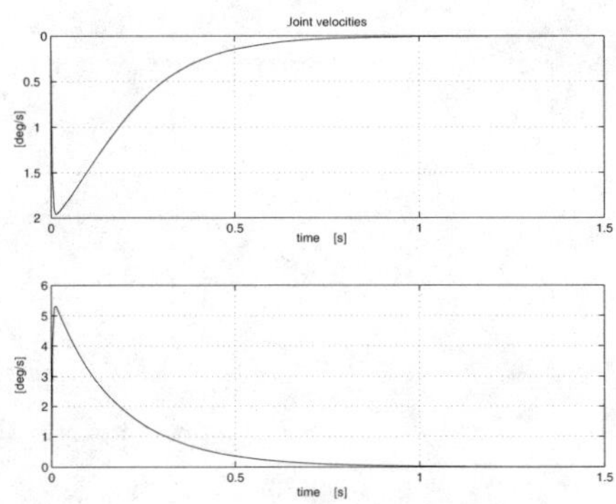

(c) Joint velocities

Fig. 2.10. Simulation results with PD + g(q).

i.e. $K_p = diag(6000, 6000)$, $K_d = diag(1100, 1100)$ while $\bar{q} = diag(0.1, 0.1)$. Also in this case, the desired configuration is reached without errors. Note the behavior of the torques: after a transient, when the errors are null, a switching behavior takes place in order to constrain the state on the desired configuration (corresponding to the minimum of H_c). The final errors in this case are $e_x = 1.1433e - 005$, $e_y = -1.9248e - 006$ (m), corresponding to $e_{q_1} = -0.0006, e_{q_2} = 0.0013$ (deg). If the robot parameters are not perfectly known, the PD + gravity compensation scheme is not able to reach the desired configuration. This case is shown in Fig. 2.13, where, as limit case, it is assumed that the parameters m_1 and m_2 are not known at all (i.e. the values $m_1 = m_2 = 0$ are assumed). As expected, the robot reaches a different final configuration and the final errors are not null: $e_x = -0.0098$, $e_y = 0.1134$ (m)

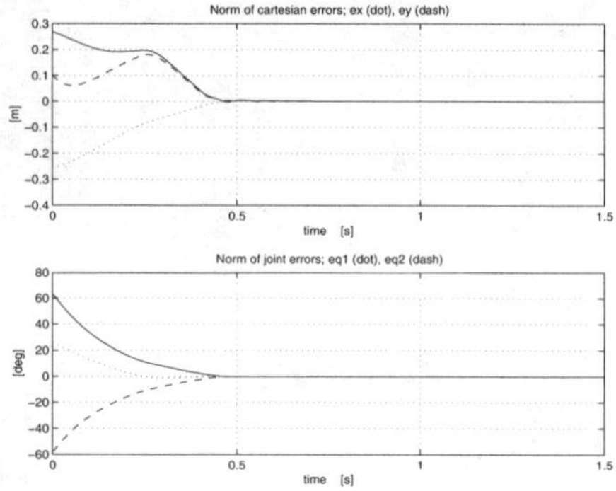

(a) Errors.

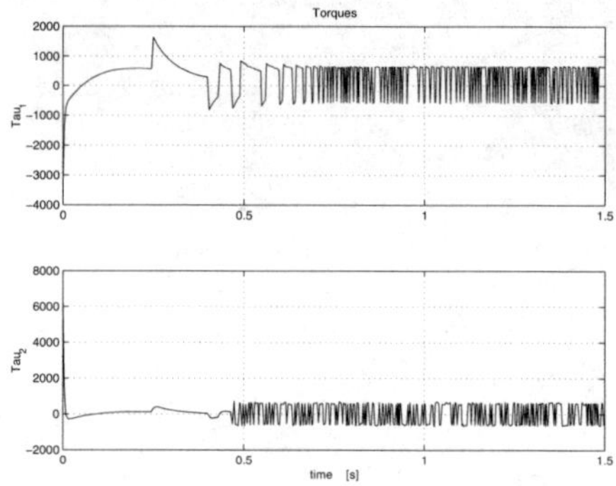

(b) Torques.

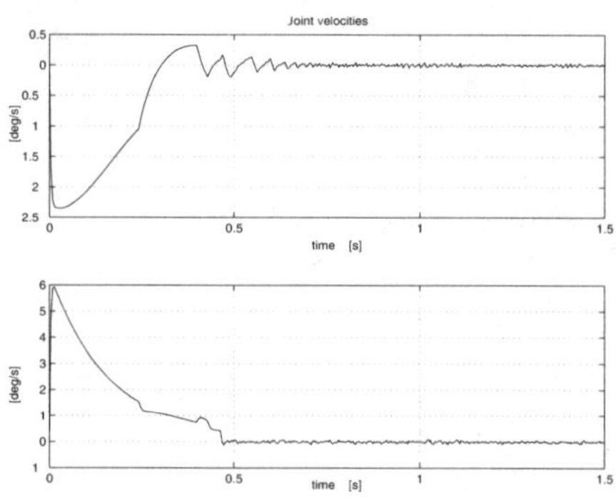

(c) Joint velocities

Fig. 2.11. Simulation results: errors, torques and velocities

$e_{q_1} = -3.2861$, $e_{q_2} = -0.84212$ (deg). The corresponding simulation with the proposed controller is shown in Fig. 2.14. Note that in this case the desired configuration is reached without errors. In this case, the final errors are $e_x = -1.2406e - 006$, $e_y = 0.000017$ (m) and $q_1 = -0.0005$, $q_2 = -0.00003$ (deg).

Finally, the case of saturation has been considered. A saturation value of $800Nm$ has been considered for the actuators. Results obtained with the proposed controller (and no knowledge of the parameters m_1 and m_2) are reported in Fig. 2.15. Errors in this case are $e_x = -9.0561e - 006$, $e_y = 1.2897e - 005$ (m) $q_1 = 0.000056$, $q_2 = -0.00098$ (deg).

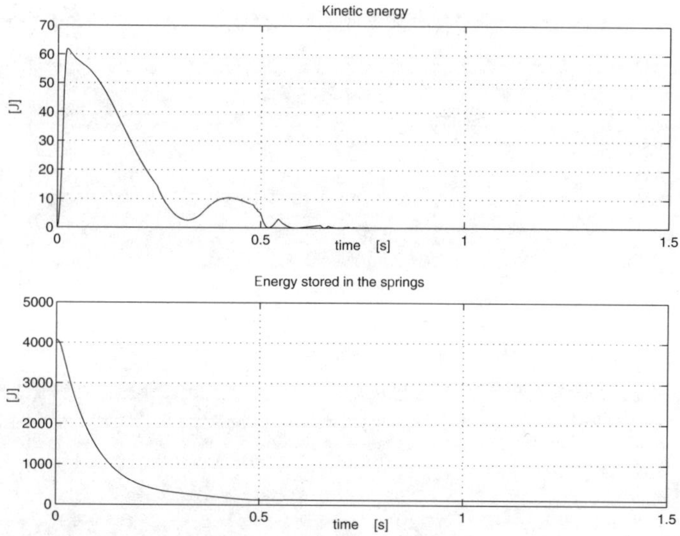

Fig. 2.12. Simulation results: energy.

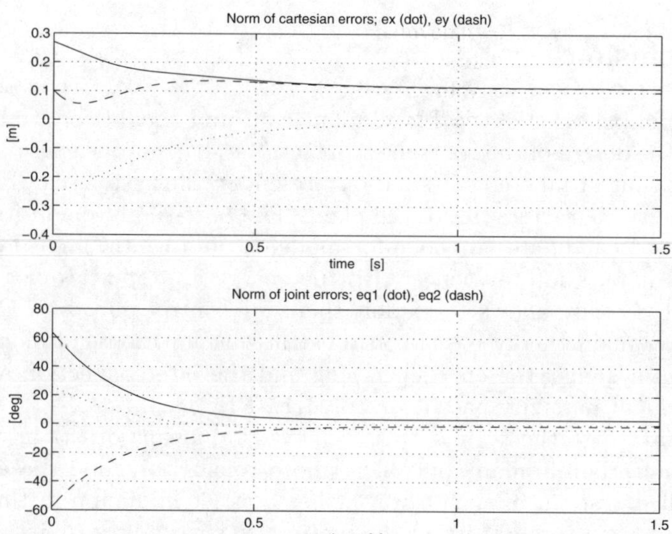

Fig. 2.13. Simulation results with PD + g(q) and partial knowledge of mass parameters: errors.

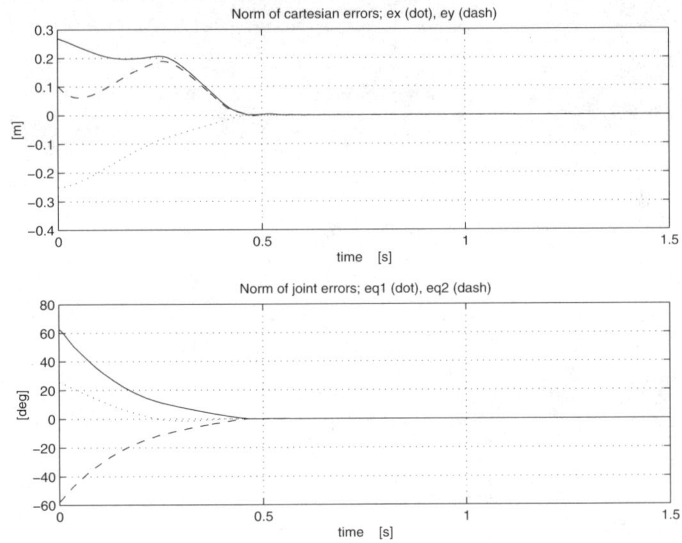

Fig. 2.14. Simulation results with partial knowledge of mass parameters: errors.

2.6 Conclusions

The energetic behavior of a physical plant can (and should) be fruitfully exploited for the design of the controller. In fact, the shape of the energy function and the amount of inherent dissipation present in the system can be related to the stability characteristics of the plant. The port-Hamiltonian framework is a very good candidate language for modeling and controlling physical systems since it puts into evidence ALL the energetic properties of a physical system and it easily allows to exploit them for control purposes. Currently, there are two main energy based control strategies for the regulation of port-Hamiltonian systems: the energy shaping and the interconnection and damping assignment passivity based control (IDA-PBC).

In the first case the regulation is achieved by means of the interconnection of a port-Hamiltonian controller whose role is to shape the energy of the controlled system, in such a way it has a strict minimum in the desired configuration, and to add further dissipation, in such a way that the desired configuration becomes asymptotically stable. Some recent applications of the energy shaping control can be found in [44, 321], where the implications of the composition of Dirac structures with the control by interconnection are investigated, in [110, 136], where the problem of controlling switched and hybrid control systems is considered, in [76, 75], where the port-Hamiltonian framework and energy shaping techniques are exploited for modeling and controlling walking robots.

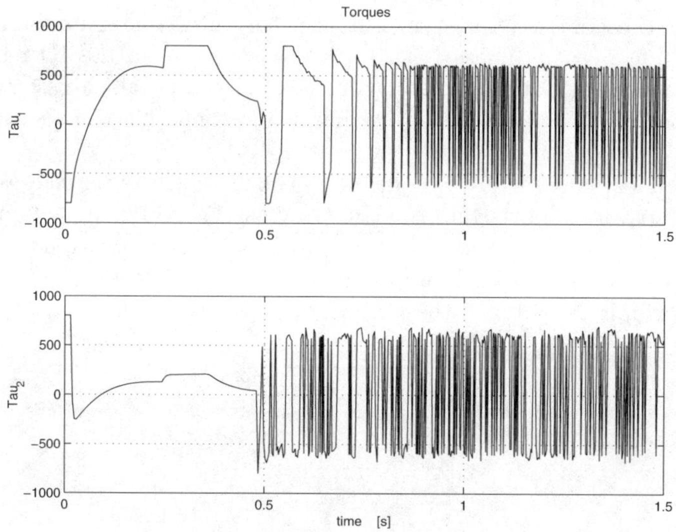

(a) Torques.

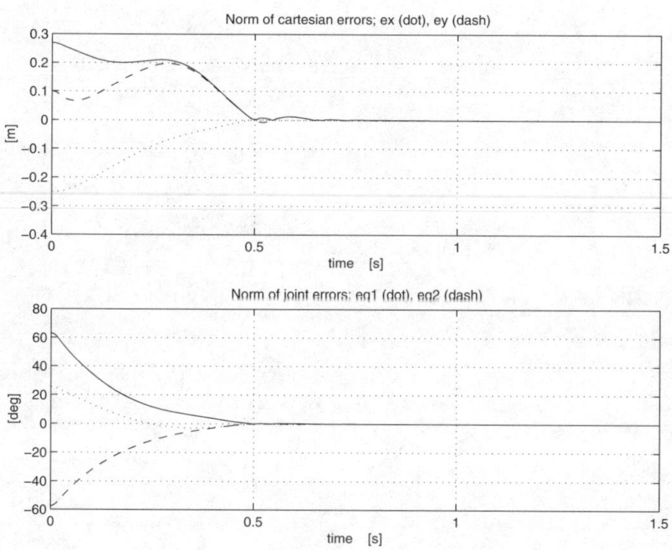

(b) Errors

Fig. 2.15. Simulation results with partial knowledge of mass parameters and saturation.

In the second case, the regulation is achieved by shaping not only the energy of the controlled system but also the interconnection structure and the damping matrix. There is still a lot of research on the IDA-PBC in several fields as, for example, in the control of electrical machines [240, 17, 99], in the control of underactuated mechanical systems [2] and also from a methodological point of view [98].

The energy shaping technique has been recently extended for the control of distributed port-Hamiltonian systems [174, 176] and of mixed port-Hamiltonian systems (namely, consisting of the interconnection of distributed and non distributed port-Hamiltonian systems) [175] by generalizing the notion of Casimir function to the infinite dimensional case.

3

A Port-Hamiltonian Approach to the Control of Interaction

3.1 Introduction

In many applications, a robot has to interact with the surrounding environment in order to perform some useful task. When a manipulator interacts with an object a very profound change occurs. In fact, before the contact, the controller has to control only the motion of the robot; after the contact, the manipulator dynamically interacts with the environment and the controller has to manage a *new* dynamical system made up by the robot coupled with the environment. It has been proven in [328] that even if the controlled robot is stable in case of free motion, its behavior could become unstable when there is a contact with the environment.

Many methodologies for the control of interaction have been developed and a lot of control algorithms have been proposed. A very successful strategy is the so called *impedance control* [122, 123] which is the starting point for the development of a port-Hamiltonian based intrinsically passive control of interaction. The main idea in impedance control is that the environment cannot be simply treated as a signal generator and, consequently, the effect of the contact of the manipulator with the environment cannot be interpreted as an exogenous disturbance. In fact, interaction is a bidirectional physical phenomenon where both the interacting systems reciprocally influence each other. For example, consider a robot touching a soft obstacle: the obstacle stops the motion of the robot and the robot deforms the obstacle. Furthermore, treating interaction as a disturbance can be very misleading for control purposes; in fact the way the robot reacts to the environment is as important as its response to a particular reference trajectory and thus, the dynamics of the contact has to be explicitly taken into account.

In robotics, and, in general, in the mechanical domain, contact takes place through localized ports through which the interacting systems influence each other. The information exchanged through the interaction ports concerns velocities and forces, by means of which the systems dynamically interact. Consider a mechanical system that may interact through its end-effector with the environment. What variable should be controlled: the velocity or the force at the end-effector? Following the impedance control paradigm, the answer is: *none of them.* In fact velocity and force at the interaction port depend both on the dynamics of the mechanical system, which is known, and on the dynamics of the environment, which is often very poorly characterized. The only thing that is independent of the dynamics of the environment is the *dynamical relation* between the force and the velocity at the interaction port. The controller, therefore, should rather control the behavior at the interaction port variables instead of a single variable. For further details on impedance control, see [122].

Port-Hamiltonian formalism can be fruitfully exploited to control the interaction of robotic systems with external, possibly unknown, environments. In this chapter it is shown how to achieve a passive interaction between a robot and any passive environment by using the Intrinsically Passive Controllers (IPC) [290] that can be modeled as port-Hamiltonian system; it is then reported how to extend the standard IPC structure used for fully actuated robots in order to increase performances when dealing with defective manipulators [271].

Furthermore a novel scheme for haptic interfaces, which allow a human operator to interact with virtual environments, is proposed. Firstly it is illustrated how to build an intrinsically passive haptic scheme which allows to interact with any virtual physical system, as reported in [297, 298]. Secondly the scheme is extended in order to take into account delayed virtual environments and the possibility of scaling the force transmitted to the user, as reported in [262, 298, 267]. Moreover the effect of quantization on sensors is taken into account and the scheme is modified in order to guarantee passivity independently of quantization errors, as illustrated in [261].

3.2 Intrinsically Passive Control of Interaction

When two systems are interacting, they can be modeled as exchanging force and velocity information through a localized port. Thus, it is possible to model the interaction port by means of a power port (see Sec. 1.3) and the reciprocal influence between the interacting systems as an exchange of energy. Furthermore, since the only thing that can be intrinsically controlled is the relationship between force and velocity at the interaction port, it is necessary to build a controller that is able to modify the *behavior* (or, in other words, the way in which energy is exchanged) of the system at the port instead of regulating only one variable.

Consider a dynamical system which can interact with the external environment through an interaction port which is characterized by a pair of power conjugated variables (e, f) (e.g. twists and wrenches for a robot). It is possible to model e and f as manifest variables of the system whose behavior depends on the particular dynamics of the system. The aim of the control of interaction is to regulate the relationship between the manifest variables, namely to control the (manifest) behavior of the system. Thus, the *control by interconnection* paradigm (see Sec. 2.3.2) can be fruitfully exploited. In fact, it is possible to regulate the behavior of the system by interconnecting to it another dynamical system, the controller, whose task is to constrain the behavior of the plant at the power port through which it is interacting with the environment (e.g. the end-effector for a robotic manipulator) to a desired subset. In this way, the relationship between the manifest variables that the system can present at the interacting port is regulated *independently* of the environment the system is in contact with.

Since the interaction between physical systems consists of an *exchange* of energy through a power port, if two passive systems are interacting, passivity, and consequently a stable behavior, is preserved even if discontinuous contacts, bouncing and other effects are present. A lot of physical systems (e.g. an n-DOF mechanism) are passive and they can be modeled as port-Hamiltonian systems. Thus a non controlled physical system preserves a passive behavior both in case of contact and non contact with any passive environment.

It is very often necessary to control the way in which the physical plant interacts (e.g. for setting a stiff or a soft interaction). In order to maintain a passive behavior it is useful to interconnect to plant a passive controller in a power preserving way. In this way the controlled plant is still passive and, therefore, passivity in the interaction with ANY, possibly unknown, passive environment is preserved. This particular kind of controller is called *Intrinsically Passive Controller* or, shortly, IPC [290] and it can be interpreted as a physical system that compensates some unwanted properties of the plant and that sets the desired behavior for the interaction.

The coupling between the IPC and the plant guarantees that the controlled system behaves in the desired way when interacting with any passive environment. In order to perform some useful tasks, it is necessary to be able to inject some energy into the controlled system. It is then necessary to endow the IPC with an additional power port through which it is possible to supply energy to the controlled plant. This leads us to the general scheme for the control of interaction that is represented in Fig. 3.1 [290].

The half-arrows indicate an energy flow in a bond-graph notation. The plant can interact, i.e. to exchange energy, with the environment. The behavior of the interaction is regulated by an intrinsically passive controller, IPC, which is connected in a power preserving way to the plant. The IPC is endowed with an extra power port through which a supervisory system can provide the energy necessary to perform a certain task to the controlled system. The port through which the supervisory system supplies energy is called *supervisory port* or, when considering the mechanical equivalent of the controller, as we will do in the following, *virtual point*.

The intrinsically passive control of interaction can be elegantly described within the port-Hamiltonian framework. In fact, the physical plant can be mo-

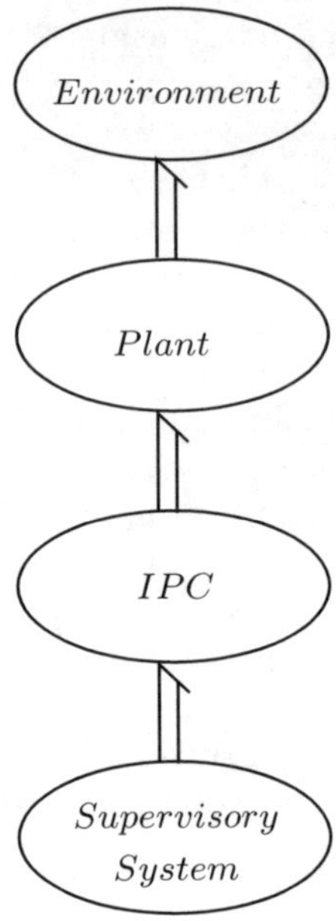

Fig. 3.1. The general form of an intrinsically passive control scheme

deled as a port-Hamiltonian system and it is interconnected to another port-Hamiltonian system, the IPC, whose role is to shape the energetic properties of the plant in such a way that the controlled system, while interacting with a passive environment, behaves as desired. Thus the design of the IPC can be done using the techniques presented in Sec. 2.3. Since the IPC is intrinsically passive, any error in the model of the plant can influence the performances of the controlled system but *never* its passivity. Thus, the controlled system will stay stable in any situation, both in case of contact and non contact with any passive, possibly unknown, environment. Performances can be increased using the variable structure energy shaping technique presented in Sec. 2.5.

3.3 Intrinsically Passive Control of Robotic Systems

The aim of this section is to illustrate two very useful applications of the IPC: the control of interaction of anthropomorphic robotic arms and the control of robotic hands.

The configuration of a robot is commonly represented by an homogeneous matrix h, a 4×4 matrix that incorporates both translational and rotational information. The set of the homogeneous matrices is the special Euclidean group, denoted by $SE(3)$ and it has the structure of a Lie group . Twists and wrenches are geometric entities that generalizes the concepts of velocity and force; in particular, twists represent a screw motion, namely an instantaneous

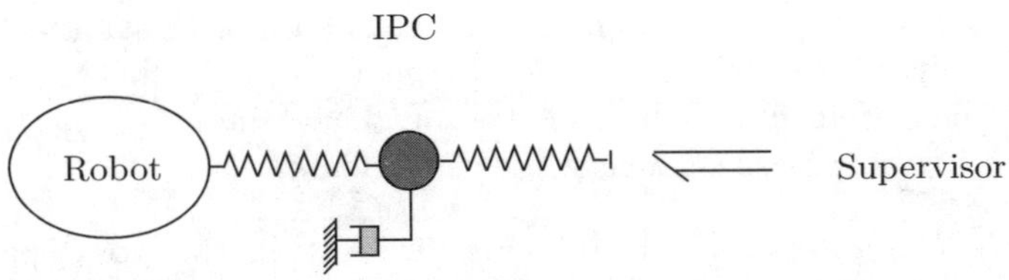

Fig. 3.2. The IPC for anthropomorphic robots

translation and an instantaneous rotation along a certain instantaneous axis in space, while wrenches represent the action of a force along a certain direction and of a moment along the same direction. This concepts are very well known in robotics; a short background on Lie groups, twists and wrenches in provided in Appendix A. Further details can be found in [212, 164, 272, 290]. Twists and wrenches are power conjugated variables, they can be modeled as flows and efforts respectively, and constitute the power port through which a robot interacts with the environment.

The following notation is used in this section:

- Ψ_i: i-th orthonormal reference frame.
- h_i^j: Homogeneous matrix representing the configuration of Ψ_j with respect to Ψ_i.
- $T_i^{k,j} = ((\omega_i^{k,j})^T, (v_i^{k,j})^T)^T$: Twist of Ψ_i with respect to Ψ_j expressed in Ψ_k.
- $Ad_{H_i^j}$: Adjoint matrix to change the representation of a twist from Ψ_i to Ψ_j.
- W_i^k: Wrench applied to a mass attached to Ψ_i expressed in Ψ_k.
- $W_{i,j}^k$: Wrench applied by a spring element connecting Ψ_i to Ψ_j, on the side of Ψ_i expressed in Ψ_k.

3.3.1 IPC for Anthropomorphic Robotic Arms

The IPC can be represented as a virtual physical system and a very well suited structure for an IPC controlling an anthropomorphic robot is the one represented in Fig. 3.2. The IPC is interconnected virtually to the end-effector of the robot and can be represented as a physical system made up of several components: a mass (called *virtual object*) and two springs, energy storing elements, and a damper, an energy dissipating element. Energy can be supplied to the system in order to perform some useful tasks and, by the IPC, we can obtain, independently of the energy supplied, a desired behavior of the interaction; in particular, we can obtain a desired compliance by means of the stiffness of the springs. Furthermore, the presence of the virtual object allows to dissipate energy by means of the damper and this leads to a strictly passive and, therefore, as reported in Sec. 2.2.3, asymptotically stable behavior of the system.

Remark 3.1. Notice that the virtual object is part of the controller and thus energy dissipation is achieved *without* velocity measurements but only with position measurements of the end-effector. Thus, damping injection is obtained by using only position sensors (e.g. encoders).

The *virtual object* is a rigid body free to move in the work space. It is an element storing kinetic energy and is therefore characterized by a function E_k expressing the stored energy. Such an element can be seen like a port-Hamiltonian system of the following form:

$$\begin{aligned} \dot{P}_b^b &= J_b \frac{\partial E_k(P_b^b)}{\partial P_b^b} + W_b^b \\ T_b^{b,0} &= \frac{\partial E_k(P_b^b)}{\partial P_b^b} \end{aligned} \tag{3.1}$$

where P_b^b is the generalized momentum of the body b expressed in a reference frame Ψ_b posed in the center of the body, W_b^b is the wrench applied on the body b expressed in Ψ_b, $T_b^{b,0}$ is the twist of the body b respect to the inertial frame Ψ_0 expressed in Ψ_b. $E_k(P_b^b)$ is the kinetic energy function and J_b is the skew-symmetric structure matrix of the body representing gyroscopic effects.

The *damper* is an element which dissipates free energy and which is characterized by the following equation:

$$W_b^b = RT_b^{b,0} \tag{3.2}$$

where R is a symmetric positive semidefinite dissipation matrix in Ψ_b. This element is not characterized by a state since it doesn't store energy.

The *springs* used in the IPC are *spatial springs*, namely springs which act on $SE(3)$ (see Appendix A) and that, therefore, can apply both torques and forces along all the directions.
A spring stores potential energy. It is characterized by a function $V_{i,j}$ representing the stored energy. This function depends on the relative position between the two bodies i and j connected by the spring. We can schematically describe a spatial spring (in 3D) as:

$$\begin{aligned} \dot{h}_i^j &= R_{h_i^j} T_i^{j,j} \\ W_{i,j}^j &= R_{h_i^j}^T \frac{\partial V_{i,j}}{\partial h_i^j} \end{aligned} \tag{3.3}$$

where h_i^j is a local coordinate of $SE(3)$, representing the relative position between the bodies i and j, and $R_{h_i^j}$ is the Lie group right translation in the coordinates h_i^j (see Appendix A); the pair $(T_i^{j,j}, W_{i,j}^j)$ is the power port associated to the deformation of the spatial spring. It is sometimes useful to consider a more general type of spatial spring: the *variable rest length spring*. This kind of spring has an additional power port by means of which it is

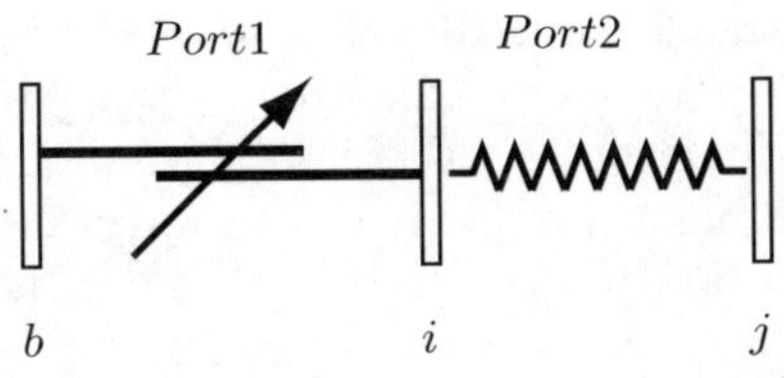

Fig. 3.3. The 1-DOF variable length spring.

possible to change its rest length and its schematic representation in the one dimensional case is illustrated in Fig. 3.3. In order to describe analytically this element, we can add to Eq.(3.3) a power port that can be used to change its rest length. For this reason, we can describe the twist associated to this spring as:

$$T_i^{j,j} = T_b^{j,j} + Ad_{h_b^j} T_i^{b,b} \tag{3.4}$$

where the first element is the usual term deriving from the deformation of the spring, while the second one takes into account the modification of the rest length. From Eq.(3.3) and Eq.(3.4), the port-Hamiltonian description of the variable rest length spring is:

$$\begin{aligned} \dot{h}_i^j &= \begin{pmatrix} R_{h_i^j} & R_{h_i^j} Ad_{h_b^j} \end{pmatrix} \begin{pmatrix} T_b^{j,j} \\ T_i^{b,b} \end{pmatrix} \\ \begin{pmatrix} W_{b,j}^j \\ W_{i,b}^b \end{pmatrix} &= \begin{pmatrix} R_{h_i^j}^T \\ Ad_{h_b^j}^T R_{h_i^j}^T \end{pmatrix} \frac{\partial V_{i,j}}{\partial h_i^j} \end{aligned} \tag{3.5}$$

Note that this port-Hamiltonian system has two power ports: the port $(T_b^{j,j}, W_{b,j}^j)$ as a normal spring, and the additional port $(T_i^{b,b}, W_{i,b}^b)$, that can be used to modify the rest length; setting $T_i^{b,b} = 0$, the rest length remains constant and the spring behaves normally. For further details on spatial springs, see [168, 290].

The basic idea of the IPC, as shown in Fig. 3.2, is that the robot's end-effector is "attached" by means of a spatial spring to the virtual object which, on its turn, is "attached" to another spatial spring by means of whose power ports a supervisor can change the state of the whole system by introducing energy in it as described in Sec. 3.2. The supervisor, by injecting energy into the IPC, moves the virtual object and the end effector of the robot, thanks to the coupling introduced by the spatial spring, follows the motions of the virtual objects and, in case of contact with the environment, the interaction is implicitly controlled in a passive manner by the IPC.

In manipulation tasks, it may happen that the kinematic configuration of a robot does not permit mobility in the whole 6-dimensional workspace, or

also that some movements are not allowed, due to constrains (e.g. during manipulation with articulated hands) [22]. Under these circumstances the robot cannot move in every direction and therefore it cannot follow the virtual object which, on the other hand, can freely rotate and translate in the workspace under the action of the supervisor.
The final result is that the system reaches a minimum of the potential energy, minimum that in general is a 'trade-off' between the translational and rotational energies stored in the springs of the IPC. This implies that the desired motion is not tracked neither in the linear nor in the rotational part.

To solve this problem the rest length of the IPC springs can be changed in such a way that, for example, the orientation constraints of the robot are directly compensated by the controller, and only the translational movements are followed. This correction can be implemented passively within the IPC by defining a proper internal interconnection between the rotational and translational springs, [271]. For this purpose, let us consider the two frames connected by the spatial spring as a frame attached to the robot (Ψ_c) and a frame attached to the virtual object (Ψ_b). Moreover, we introduce a third reference frame (Ψ_v) attached at the point i to take into account the variability of the spring rest length, see Fig. 3.3. With reference to these frames, a motion of the virtual object is described by the twist $T_b^{0,0}$, where Ψ_0 is an inertial frame.
Given a desired twist $T_c^{c,0}$ of the robot, it is well known that the following relationship holds

$$\begin{pmatrix} \omega_c^{c,0} \\ v_c^{c,0} \end{pmatrix} = T_c^{c,0} = J(q)\dot{q} = \begin{pmatrix} J^\omega(q) \\ J^v(q) \end{pmatrix} \dot{q}$$

being $q, \dot{q}$ the joint position and velocity vectors, $J(q)$ the Jacobian matrix and $J^\omega(q), J^v(q)$ the sub-matrices related to the rotational and linear velocity respectively.
Given a twist $T_b^{0,0}$ imposed to the virtual object, in which in general $\omega_b^{0,0} \neq 0$, we would like the controller to generate only a translational velocity of the end-effector computed as:

$$v_d^{0,0} = Q T_b^{0,0}$$

where $Q = [0_3,\ I_3]$, where 0_3 and I_3 represent the 3×3 zero matrix and identity matrix respectively, is used to extract the translational velocity from the twist. Since

$$v_c^{0,0} = R_c^0 J^v \dot{q} = K\dot{q}$$

being R_c^0 the rotational matrix from Ψ_c to Ψ_0, we obtain the joint velocity required for tracking $v_d^{0,0}$ as:

$$\dot{q} = K^+ v_d^{0,0} = K^+ Q T_b^{0,0}$$

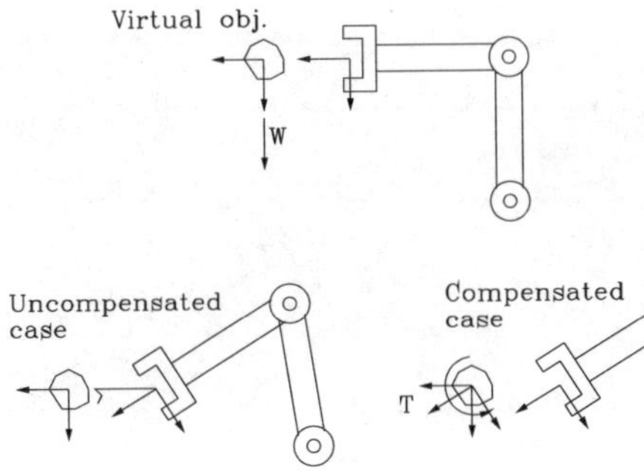

Fig. 3.4. A two DOF defective robot

where K^+ is the pseudo-inverse of matrix K, defined if proper metrics are given in the joint and task spaces. Then, the resulting twist is computed as

$$T_c^{c,0} = JK^+QT_b^{0,0}$$

We want to act on the length port of the spring (therefore on $T_v^{b,b}$) in order to have $T_v^{c,0} = T_c^{c,0}$, i.e. the twist of the frame on the length port must be achievable by the robot. Since in general

$$T_v^{b,b} = Ad_{h_c^b}T_v^{c,0} - Ad_{h_0^b}T_b^{0,0}$$

by imposing $T_v^{c,0} = T_c^{c,0}$ we can write

$$T_v^{b,b} = (Ad_{h_c^b}JK^+Q - Ad_{h_0^b})T_b^{0,0}$$

which is the input to be given to the power port of the spring in order to compensate for the rotational velocity of the virtual object.
By using an IPC with standard springs, the energy deriving from the motion of the virtual object goes into the springs and changes both the position and the orientation of Ψ_c of the robot. The spring, since it is a 3D spring, tries to align the two frames from both the rotational and the translational point of view. By using a variable spring, part of the energy deriving from the motion of the virtual object is deviated towards the length port, and the desired translational motion can be followed by the robot.

As a simple case study, the above technique has been applied to the control of a two DOF defective robot, shown in Fig. 3.4. The desired task is a motion in the vertical direction, imposed by applying a vertical force W to the virtual mass. For the sake of clearness, in Fig. 3.4 both the uncompensated and compensated case are shown. In the first case the robot tries to follow the vertical motion of the virtual object and to maintain the initial orientation. Since this is not possible, the rotational part of the spring is compressed and the final configuration is not the desired one. In the second case, the change of orientation is taken into account directly by the IPC and the robot may reach

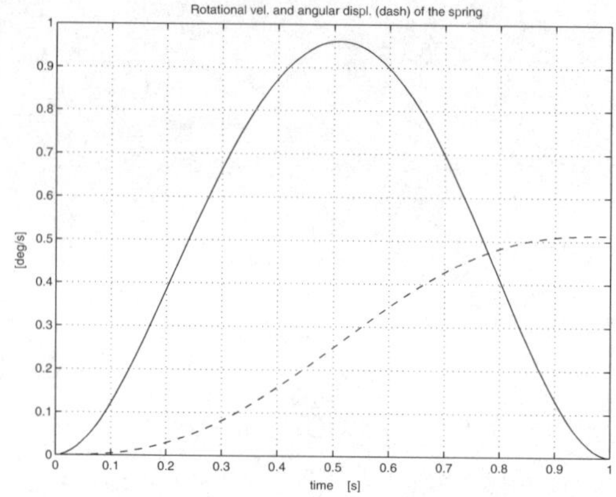

Fig. 3.5. Rotational velocity $\omega_v^{b,b}$ and displacement (dashed) at the the variable length port of the spatial spring.

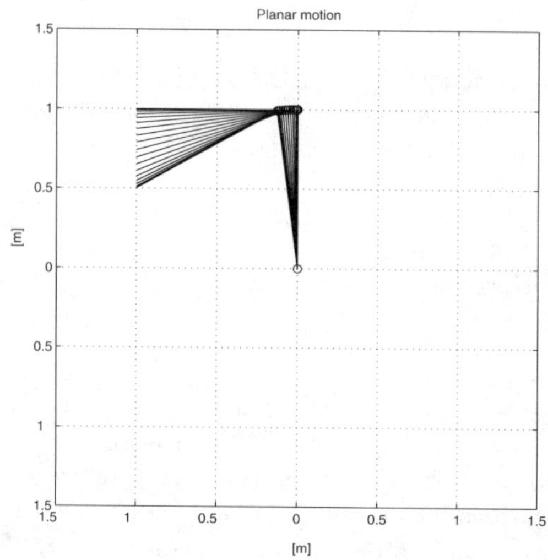

Fig. 3.6. Desired motion of the planar robot.

the desired position. This result is shown also in Fig. 3.5, where the rotational velocity and the resulting angular displacement (proportional to the torque generated in the uncompensated case) are reported. The motion of the robot is shown in Fig. 3.6.

3.3.2 IPC for Grasping

Robotic hands have to interact with the environment in order to grasp objects and to manipulate them. The IPC, therefore, can be profitably applied to these devices in order to control the interaction of the fingers with the objects to

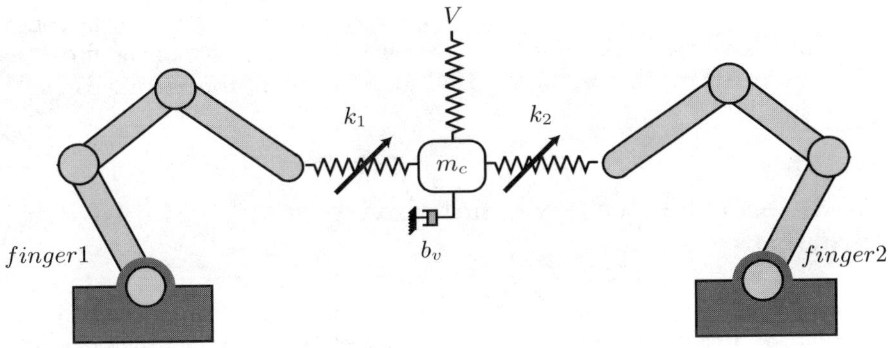

Fig. 3.7. The grasping system

be grasped. On the other hand it is necessary also to freely move the fingers in the workspace in order to approach the object to be manipulated. It is possible to build an IPC that can both regulate the contact of the fingers with an object and allow the motion of the fingers. The control scheme is represented in Fig. 3.7 for the simple case of an hand with two fingers.

The arrows on the springs express the fact that they are variable rest length springs. Each fingertip is connected to the virtual object by a variable rest length spring and the virtual object is connected to the virtual point by means of a spatial spring. The supervisory system can inject energy into the system both through the virtual point and through the rest length ports of the spatial springs that connects the fingertips to the virtual object. When some energy is supplied through the rest length port of the robotic hand, the rest lengths of the spatial springs connecting the fingertips to the virtual object change and, therefore, the relative position of the fingers with respect to the virtual object changes. In this way it is possible to change the relative position of the fingers in the space. When some energy is provided through the virtual point, the position of the virtual object changes and, therefore, the position of all the fingers in space changes.

A simple grasping strategy can be implemented by using the proposed IPC:

1. Supply energy through the rest length ports in order to open the fingers
2. Move the virtual object in the proximity of the object to be grasped moving the virtual point
3. Change the rest length ports in order to close the fingers to grasp the object

The contact is controlled by means of the IPC which allows, through a proper choice of the stiffness of the spatial springs connecting the fingertips to the virtual object, to choose the compliance of the contact. Furthermore the presence of the virtual object allows to dissipate energy, and, therefore, to obtain a strictly passive (i.e. asymptotically stable) behavior of the robotic hand by dissipating energy without measuring the velocity of the fingertips. Further details on the use of the IPC for controlling robotic hands can be found in [290].

3.4 Interaction with Virtual Environments: Haptic Interfaces

An haptic interface is basically a system which allows a human operator to interact, by means of some robotic device, with a virtual environment simulated at a certain sampling rate. As in the implementation of every control algorithm for interaction tasks, stability is a key issue since either oscillations or unstable behaviors can lead to an unnatural feedback from the virtual environment or even to harmful situations for the human operator. The stability analysis of the haptic device is not a trivial task and is hardly solvable with parameter based tools of non-linear control. As a matter of fact, interesting virtual environments are very often non-linear and the dynamic of the human operator, which has a non-negligible role in the haptic chain, is difficult to model. A very suitable tool to ensure the stability of the haptic display is passivity theory. In fact, since an haptic system is nothing else than an interactive robotic interface, it is sufficient to guarantee the passivity of a system in order to have a stable behavior [319]. Furthermore, Hogan [123] has shown that the behavior of the human operator is passive in the range of frequencies of interest in haptics and, finally, in a passivity based analysis we can consider both linear and nonlinear virtual environments and haptic devices.

A rigorous examination of the stability of an haptic display has been made in [59] where it is shown which is the necessary amount of damping to ensure the passivity of the system, once a model of the virtual environment and the sampling time are given; recently, this kind of stability analysis has been extended and improved in [71]. Colgate introduced the idea of *virtual coupling*[60] that makes possible to guarantee passivity for arbitrary passive operators and environments and even for a class of non passive environments [202]. Roughly speaking, the virtual coupling is a virtual mechanical system which is interposed as a layer between the haptic interface and the virtual environment to limit the maximum or the minimum impedance presented by the virtual environment in such a way to guarantee stability. Adams [5, 3, 4] developed a method based on two-port theory and Llewellyn's criterion for absolute stability to design the virtual coupling for both impedance and admittance causalities of the haptic display.

A possible drawback of the fixed parameters virtual coupling lies in performance decreasing, which is related to an excessive energy dissipation in some working condition. In [114, 115, 250] Hannaford introduced a virtual coupling strategy with variable parameters, called PO/PC. This strategy implements, loosely speaking, a variable damper that is activated only when an energy

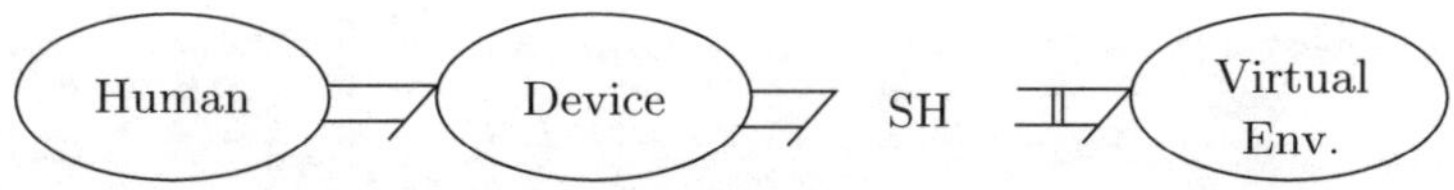

Fig. 3.8. Energetic representation of a haptic display

increment is detected and therefore the presence of this *variable virtual layer* affects the performances of the haptic interface as little as possible.

An haptic interface consists of a human operator that exploits an haptic device by means of which he can interact with a discrete-time virtual environment. Thus, we can model the haptic display as an energetic interconnection of systems as shown in Fig. 3.8. This scheme shows the energetic exchange among the system components by means of the bond-graph formalism [237]. The bond with the double vertical bar indicates an exchange of energy which occurs in discrete-time, i.e. a virtual exchange of energy. The element denoted with SH represents the Sample & Hold component, which implements the gate between the continuous and discrete domains. The continuous output power variable is sampled and it becomes the discrete input of the discrete virtual environment. The discrete power conjugated output of the virtual environment is held and given as an input to the haptic device.

In the scheme there are some points in which we can have energy generation instead of simple energy exchange. This phenomenon leads to a non passive behavior and, therefore, to a possibly unstable behavior of the haptic device. The factors which contribute to the production of extra energy have been called "energy leaks" [102] and two main energy leaks can be distinguished in an haptic interface:

- Zero Order Hold
- Discrete Virtual Environment

The zero order hold can represent an energy leak because it keeps a power variable to a constant value during the sample period, regardless of the actual value of its conjugate variable whose behavior could be such to introduce extra energy into the system. The implementation of the virtual environment can also represent a problem. In fact, passivity is not preserved by standard discretization algorithms and, therefore, it can very likely happen that a virtual environment, obtained by discretizing its continuous passive counterpart, is not passive.

Example 3.2. An ideal physical linear spring is a lossless system and, therefore, if some energy is stored into the spring by squeezing it, then the same amount of energy will be removed by releasing it.
Consider a virtual spring implemented using Euler integration method, namely:

$$x(k+1) = x(k) + Tf(k) \tag{3.6}$$

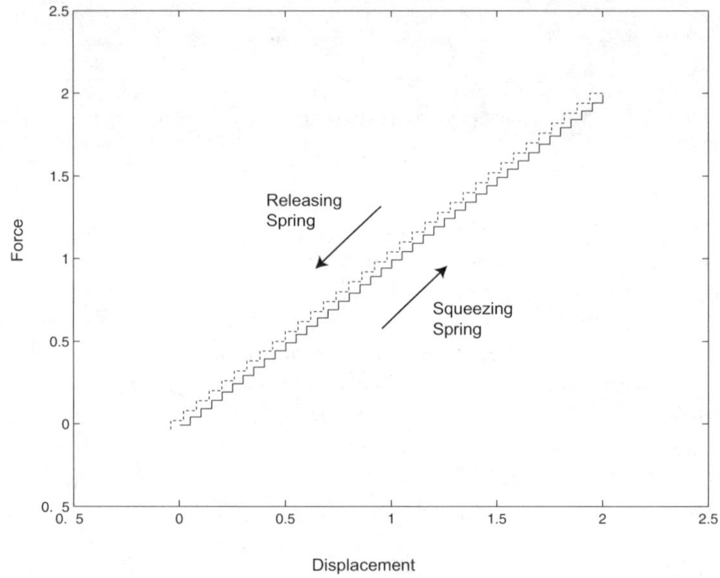

Fig. 3.9. Active behavior of a virtual spring

where x represents the displacement of the virtual spring, T the sampling period and $f(k)$ the discrete flow.
Since the spring is implemented in discrete time, the force does not increase smoothly but it is "held" at a constant value at each sample period. In Fig. 3.9 it is illustrated the behavior of a virtual spring for a squeeze/release operation. The solid and the dashed lines represent the force provided by the virtual spring versus the displacement in case of squeezing and of releasing the spring. While in case of continuous springs the two lines would coincide, for the virtual spring the average force during release is slightly greater. Thus, the virtual spring is not lossless anymore but some extra energy is generated during the release process.

Thus, the discretization of a passive system can lead to a non passive virtual system and, therefore, the discrete virtual environment must be designed carefully in order to avoid non passive behaviors in the haptic interface.

The aim of this section is to develop an intrinsically passive haptic interface based on port-Hamiltonian systems. In this way, we can model very generic haptic interfaces; in fact we can describe any kind of physical haptic device and any kind of environment to be simulated, both linear and nonlinear, as port-Hamiltonian systems. Furthermore, the port-Hamiltonian structure allows to make a very accurate energetic analysis of the haptic interface and of its energy leaks. In order to obtain an intrinsically passive haptic interface, we want to implement each possible source of extra energy passively. Thus, we firstly seek for a discretization method for port-Hamiltonian systems that preserves their passivity property. In this way, we will be able to simulate any kind of passive environment without introducing extra energy into the haptic interface [297, 298]. Secondly, we make an energetic analysis of the Sample & Hold device and we develop an energetically consistent (and therefore without any extra energy production) interconnection between continuous and discrete port-Hamiltonians [297, 298]. In this way it is possible to interconnect the haptic device with the virtual environment without violating passivity. Once every energy leak has been implemented passively, the overall haptic interface is intrinsically passive and, therefore, characterized by a stable behavior. This method minimizes the need of acting on the system through extra dissipating dynamics, increasing thus the transparency of the haptic interface and the reliability of the feedback perceived by the user.

In the following, we will consider the case in which the virtual environment has an impedance causality (flow in/effort out). This case is very common in practice since usually the human operator moves a virtual pointer and perceives the virtual world through a force feedback. Nevertheless, all the results that will be presented can be straightforwardly extended to the case of virtual environments with an admittance causality (effort in/flow out).

3.4.1 Sampled Port-Hamiltonian Systems

Consider a generic implicit port-Hamiltonian system. Once a coordinate frame has been fixed on the state manifold, it can be represented (see Sec. 1.4.1) as:

$$F(x)\begin{pmatrix} f_C \\ f_R \\ f_I \end{pmatrix} + E(x)\begin{pmatrix} e_C \\ e_R \\ e_I \end{pmatrix} = 0 \tag{3.7}$$

where subscript I indicates the power ports by means of which the system interacts with the rest of the world, with the subscript C the power ports associated with the storage of energy and with the subscript R the power ports relative to the dissipative part. If $E(x)$ is invertible we represent a port Hamiltonian system by

$$\begin{pmatrix} e_C \\ e_R \\ e_I \end{pmatrix} = -E^{-1}(x)F(x)\begin{pmatrix} f_C \\ f_R \\ f_I \end{pmatrix} = D(x)\begin{pmatrix} f_C \\ f_R \\ f_I \end{pmatrix} \tag{3.8}$$

$D(x)$ is a skew-symmetric matrix. In fact, since the matrices $E(x)$ and $F(x)$ are a Kernel representation of a Dirac structure, we have that Eq.(1.47) holds and that, therefore

$$F^T(x) = -E^{-1}(x)F(x)E^T(x)$$

Consequently

$$D^T(x) = [-E^{-1}(x)F(x)]^T = -F^T(x)E^{-T}(x) =$$

$$= E^{-1}(x)F(x)E^T(x)E^{-T}(x) = -D(x)$$

Thus, we have that the port-Hamiltonian system can be represented by

$$\begin{pmatrix} e_C \\ e_R \\ e_I \end{pmatrix} = \underbrace{\begin{pmatrix} D_C(x) & G_1(x) & G_2(x) \\ -G_1^T(x) & D_R(x) & G_3(x) \\ -G_2^T(x) & -G_3^T(x) & D_I(x) \end{pmatrix}}_{D(x)} \begin{pmatrix} f_C \\ f_R \\ f_I \end{pmatrix} \tag{3.9}$$

where $D_I(x), D_C(x), D_R(x)$ are skew-symmetric; the dependency of the sub-matrices from x will be omitted to lighten the notation. Due to the power continuity of the Dirac structure associated to the port-Hamiltonian system, we clearly have, using coordinates:

$$P_I + P_C + P_R := e_I^T f_I + e_C^T f_C + e_R^T f_R = 0 \tag{3.10}$$

which is a power balance meaning that the total power coming out of the network structure should be always equal to zero. Furthermore the following relations hold:

$$f_C = \dot{x} \quad e_C = \frac{\partial H}{\partial x} \tag{3.11}$$

where $H : \mathcal{X} \to \mathbb{R}$ is the lower bounded Hamiltonian function. The inherent dissipation present in the system can be modeled using

$$e_R = R(x) f_R \tag{3.12}$$

as characteristic equation, with $R(x)$ positive semidefinite matrix.

Remark 3.3. All the consideration of this subsection keep on holding even considering more general dissipation structure and if the matrix $E(x)$ is not invertible but the computations would be much more tedious.

After having included Eq.(3.12) in Eq.(3.7), it is possible to see, after very lengthy but straightforward calculations, that it is possible to represent the port-Hamiltonian system with dissipation with the following equations:

$$\begin{pmatrix} e_I \\ f_C \end{pmatrix} = \begin{pmatrix} B(x) & A(x) \\ C(x) & D(x) \end{pmatrix} \begin{pmatrix} f_I \\ e_C \end{pmatrix} \tag{3.13}$$

where

$$A = -[G_2^T + G_3^T(D_R - R)^{-1}G_1^T][D_C + G_1(D_R - R)^{-1}G_1^T]^{-1}$$

$$\begin{aligned} B = {} & G_3^T(D_R - R)^{-1}G_3 + D_I - [G_2^T + G_3^T(D_R - R)^{-1}G_1^T]* \\ & *[D_C + G_1(D_R - R)^{-1}G_1^T]^{-1}[G_1(D_R - R)^{-1}G_3 - G_2] \end{aligned}$$

$$C = [D_C + G_1(D_R - R)^{-1}G_1^T]^{-1}[G_1(D_R - R)^{-1}G_3 - G_2]$$

$$D = [D_C + G_1(D_R - R)^{-1}G_1^T]^{-1}$$

Recalling Eq.(3.11) and Eq.(3.12), the power balance in Eq.(3.10) can be rewritten as:

$$\dot{H}(x) + f_R^T R(x) f_R = -e_I^T f_I = P$$

which clearly says that the supplied power $-e_I^T f_I$ equals the increase of internal energy plus the dissipated one[1].

We can describe a discrete time port-Hamiltonian system as a continuous time port-Hamiltonian system in which the port variables are frozen for a sample interval T. In what follows we indicate with $v(k)$ the value of the discrete variable $v(t)$ corresponding to the interval $t \in [kT, (k+1)T]$. If we rewrite Eq.(3.10) for the discrete case, we have:

$$e_I^T(k) f_I(k) + e_C^T(k) f_C(k) + e_R^T(k) f_R(k) = 0 \tag{3.14}$$

Furthermore, during the interval k, we have to consider a constant state $x(k)$ corresponding to the continuous time state $x(t)$. This implies that during the interval k, the dissipated energy will be equal to $T f_R^T(k) R(x(k)) f_R(k)$ and the supplied energy will be equal to $-T e_I^T(k) f_I(k)$. In order to be consistent with the energy flows, and as a consequence conserve passivity, we need a jump in internal energy $\Delta H(k)$ from instant kT to instant $(k+1)T$ such that:

$$\Delta H(k) = -T f_R^T(k) R(x(k)) f_R(k) - T e_I^T(k) f_I(k)$$

This implies that the new discrete state $x(k+1)$ should belong to an energetic level such that:

$$H(x(k+1)) = H(x(k)) + \Delta H(k)$$

We can indicate the set of possible energetically consistent states as

$$I_{k+1} := \{x \in \mathcal{X} \quad s.t. \quad H(x) = H(x(k)) + \Delta H(k)\}.$$

Furthermore, from the discrete equivalent of Eq.(3.13), we have that:

$$f_C(k) = C(x(k)) f_I(k) + D(x(k)) e_C(k) \tag{3.15}$$

and therefore, for consistency with the continuous dynamics in which $f_C(t) = \dot{x}(t)$, the next state $x(k+1)$ should be such that:

$$f_C(k) = \lim_{T \to 0} \frac{x(k+1) - x(k)}{T} \tag{3.16}$$

where we considered the definition of the right derivative.

Remark 3.4. It is important to notice that the choice of the next discrete state, which is taken right after sampling instant $k+1$, can be done causally based on the energy exchange that has taken place in the last interval between the sampling instants k and $k+1$.

[1] $e_I^T f_I$ represents the power supplied by the system and, therefore, $-e_I^T f_I$ represents the power supplied to the system

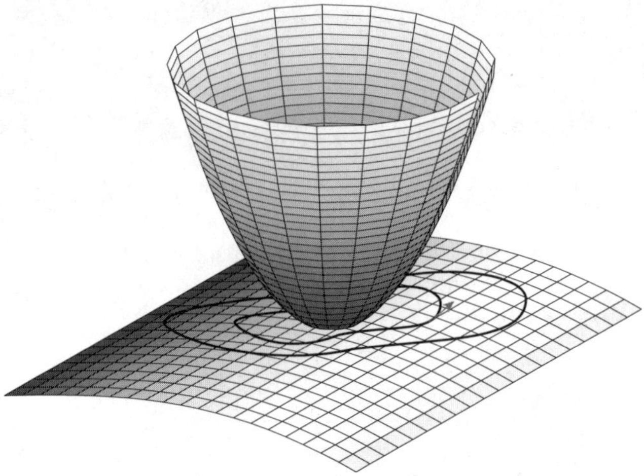

Fig. 3.10. The passivity-preserving energetic behavior of a discrete port-Hamiltonian system

The set I_{k+1} can be either empty or have more solutions.

Case in which $I_{k+1} \neq \emptyset$

This situation is the most common and corresponds to the usual one. A graphical representation is shown in Fig. 3.10 where two energy levels corresponding to the plotted energy function are given. In this case a choice should be made among the possible states of I_{k+1}. Clearly, the state should be in some sense 'close' to the current state $x(k)$ and such that the condition of Eq.(3.16) is satisfied. A picture which shows graphically the basic idea is reported in Fig. 3.11. The possible curves going through $x(k)$ and having as a tangent $f_C(k) \in T_{x(k)}\mathcal{X}$, could be characterized as geodesics once an affine connection would be defined on $\mathcal{X}$ which are indicated as dotted lines in the figure. These lines intersect the locus of states corresponding to the consistent energy set in different possible points which depend on the connection chosen. We have considered Euclidean coordinates and an Euclidean connection, which is a meaningful choice in many cases. In this case, the next state $x(k+1)$ is chosen as the intersection of I_{k+1} with the straight line passing from $x(k)$ and directed along $f_C(k)$.

Case in which $I_{k+1} = \emptyset$ and energy leap

This can happen in two situations: a) required decrease of energy close to a local minimum b) required increase of energy close to a local maximum. The case b) does not preclude passivity and therefore it will not be analyzed. In the situation a) let us indicate with x_{min} one of the states for which the energy

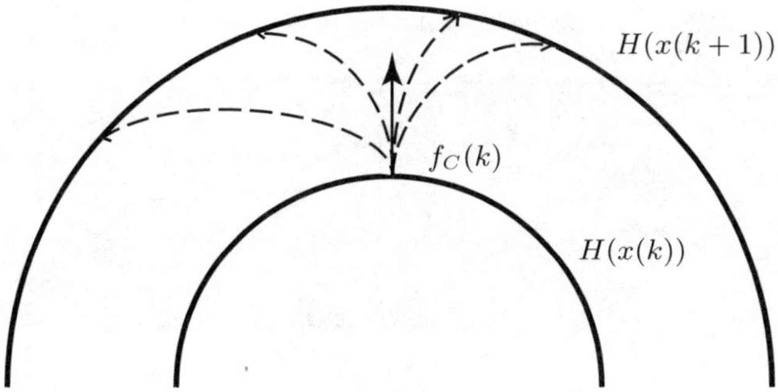

Fig. 3.11. The various possible connections for the state jump

has locally a minimum close to $x(k)$ and equal to $H(x_{min})$. This situation is therefore obtained if:

$$\Delta H(k) < H(x_{min}) - H(x(k)).$$

In this situation, it is clearly not possible to find a state $x(k+1)$ compatible with the energy change $\Delta H(k)$ since $I_{k+1} = \emptyset$. If we chose $x(k+1) = x_{min}$ we would implement the smallest error in the energy change, but this would not be a good choice for two main reasons: first, this could create a 'dynamic dead-lock' since, in this situation the effort generated by the energy function and equal to $e_C = \frac{\partial H}{\partial x}$ would be equal to zero and in case no damping would be present, it would be possible to see that this would prevent any further supply of energy from the interconnection port (e_I, f_I) since e_I could be equal to zero and therefore any further change of the state would be impossible. Second it would not help the system to behave in such a way that its dynamic makes possible to correct the energetic discrepancy due to the fact that the required $\Delta H(k)$ cannot be performed. A solution to these two problems can be found in what we call *energy leap*, which is illustrated in Fig. 3.12. Instead of choosing as a new state x_{min}, we choose as a state $x(k+1)$, a state with the same energy level, but "symmetrically positioned" with respect to the point of minimum energy x_{min}. This rather fuzzy statement could be made precise once an affine connection would be defined on the state manifold. As already said, considering Euclidean coordinates, it is possible to define the next state as the state having the same energy and lying on a straight line passing through x_{min} and $x(k)$.

Clearly, by construction, we obtain an error in the energy change equal to $\Delta H(k)$ which corresponds to the amount of energy which we supplied to the "rest of the world" through the power port (e_I, f_I). On the other hand, by the change of sign in the gradient of the energy function, we practically passed through the minimum in one go and the system will therefore now try to absorb energy from the port (e_I, f_I). In order to recover passivity it is

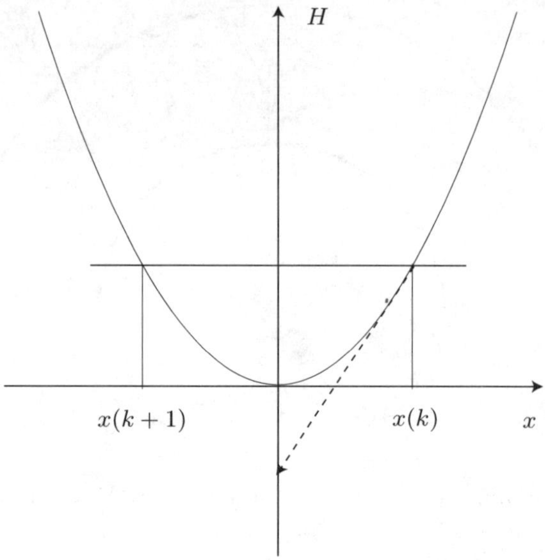

Fig. 3.12. The energy leap strategy

necessary to dissipate the extra energy introduced through the energy leap. Since we know EXACTLY the amount of energy produced we can keep track of the extra energy introduced into the system. To recover passivity we have to dissipate the extra energy produced in a *finite* number of steps; furthermore, usually, there is energy production only for one step. At each step the system makes an energy jump and most of them are passive jumps, namely:

$$H(x(k+1)) = H(x(k)) + Te^T(k)f(k) = H(x(k)) + j(k) \tag{3.17}$$

Suppose that $Te^T(k)f(k) > 0$, we can act on the magnitude of the jump in order to dissipate the energy produced: we won't, therefore,update the state in order to satisfy (3.17) but we'll introduce some dissipation by modulating the jump. The situation is illustrated in figure 3.13: Instead of jumping from $H(k)$ to $H(k+1)$ we constrain the system to jump only a fraction $0 < \alpha < 1$ of the jump $j(k)$. We reach, therefore, the energy level $H_b(k+1) = H(k) + \alpha j(k) < H(k+1) = H(k) + j(k)$ and we dissipate the quantity $H(k+1) - H(k) = (1-\alpha)j(k)$. The major point is that it is possible to keep track exactly of the error in the energy values in such a way that they can be then compensated. The closer is α to 1 the slower the "passivity recovery" is and the less influence will have the dynamics. Similar reasonings can be made in case $j(k) < 0$. This strategy for recovering passivity is called *energy book-keeping* [297, 256, 298].

As a summary of the procedure just outlined, we hereafter algorithmically explain the way the discrete system can be integrated

1. Given an initial state $x(k)$, we set $e_C(k) = \frac{\partial H}{\partial x}(x(k))$.

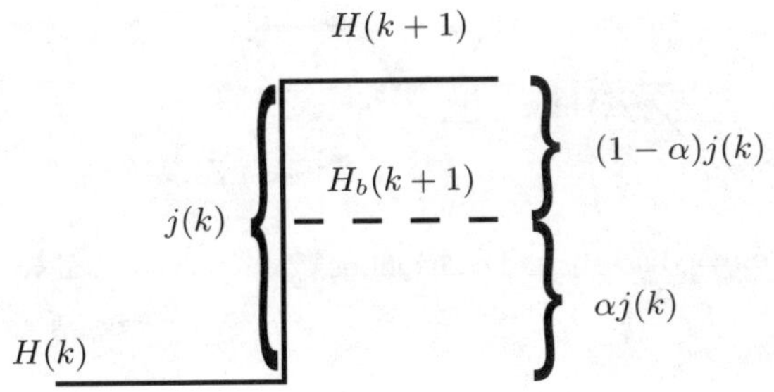

Fig. 3.13. Energy Booking for a Forward Jump

2. Using the value of the system input $f_I(k)$ and the previously calculated $e_C(k)$, we can calculate $e_I(k)$, the output of the interaction port, and $f_C(k)$ using the discrete representation of Eq.(3.13)
3. $f_C(k)$ is then used to calculate the next state $x(k+1)$ using the procedure explained at the beginning of this subsection.

3.4.2 Energy Consistent Sampled Passivity

Consider the port interconnection of a continuous time port Hamiltonian system $\mathcal{H}_C$ and a discrete port Hamiltonian system $\mathcal{H}_D$ through a sampler and zero-order hold as shown in Fig. 3.14. Suppose that $\mathcal{H}_C$ has an admittance causality (effort in/flow out) and therefore $\mathcal{H}_D$ has an impedance causality (flow in/effort out).

During the dynamic evolution of the two systems between time kT and $(k+1)T$, where T is the sampling time and k is a positive integer, the effort supplied to $\mathcal{H}_C$ by $\mathcal{H}_D$ will be constant due to the zero-order hold assumption. We will indicate this value as $e_d(k+1)$. If we indicate the power port at the continuous side with $(e(t), f(t))$, we clearly have:

$$e(t) = e_d(k+1) \qquad t \in [kT, (k+1)T]$$

By looking at the energy flow toward the continuous system, we can see that, if we indicate with $\Delta H_c(k+1)$ the energy which flows through the continuous power port from time kT up to time $(k+1)T$, we obtain:

$$\Delta H_c(k+1) = \int_{kT}^{(k+1)T} e_d^T(k+1) f(s)ds = e_d^T(k+1) \int_{kT}^{(k+1)T} f(s)ds =$$
$$= e_d^T(k+1)\left(q((k+1)T) - q(kT)\right) \tag{3.18}$$

where we indicated with $q(\cdot)$ the integral of the continuous time flow $f(t)$.

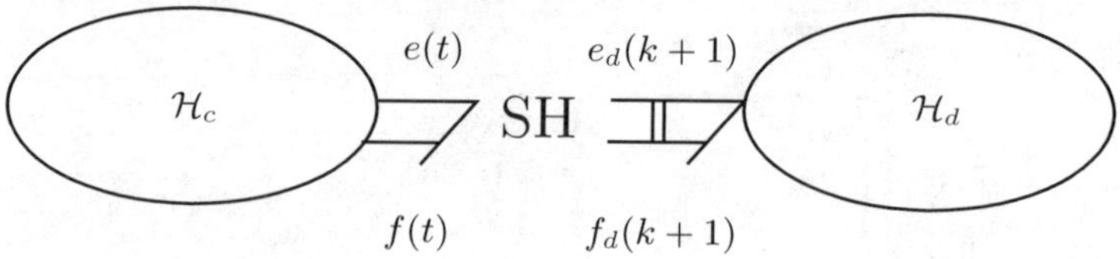

Fig. 3.14. The interconnection of discrete and continuous port-Hamiltonian systems

Remark 3.5. It is important to realize that, in most of useful mechanical applications like haptics, $e_d(k)$ will correspond to forces/torques that a controller would apply to an inertial element. In this case, $q(\cdot)$ would be nothing else than a position measurement of the masses the controller pushes on.

It is now straightforward to state the following theorem [297, 256, 298]:

Theorem 3.6 (Sampled Data passivity). *If in the situation sketched before, we define for the interconnection port of* $\mathcal{H}_D$

$$f_d(k+1) := -\frac{q((k+1)T) - q((k)T)}{T}, \tag{3.19}$$

we obtain an equivalence between the continuous time and discrete time energy flow in the sense that for each n:

$$\sum_{i=0}^{n-1} e_d^T(i+1) f_d(i+1) T = -\int_0^{nT} e^T(s) f(s) ds \tag{3.20}$$

Remark 3.7. It is important to notice that the exact equivalence is achieved only by the definition of Eq.(3.19) in which $q(\cdot)$ is usually the easiest variable to be measured in real applications. The negative sign appearing in Eq.(3.20) is consistent with the fact that the power flowing into the continuous system is minus the power flowing into the discrete side.

It may happen that during the inter-sample, passivity of the controller is lost without that the controller would find this out. This could be due to a sequence of a very high flow of energy toward the plant followed by the same amount of energy back to the controller. This would be a problem because the net energy flow towards the plant after one sample would be zero, but between the first and the second peak, there would be a moment in which the controller would have provided more energy that it had available. This "loss of passivity" between the two spikes would never be seen by the controller since this situation takes place between two samples. This situation is unavoidable because of the intrinsic loss of information due to sampling if no hypothesis is made on continuity or bandwidth, but at each sample instant the controller would adjust the energy balance as explained in the previous subsection.

It is now possible to propose the algorithm which can be used to integrate the discrete system on line. We consider to reach time $(k+1)T$ after a sample period in which the effort held on the device has been $e_d(k+1)$. After $q(k+1)$ has been measured, $f_d(k+1)$ can be calculated using Eq.(3.19). If we consider the situation at the previous time k and define $f_I(k) := f_d(k+1)$, we are able to calculate the state $x(k+1)$ by exactly following the algorithm of the previous subsection and the value $e_I(k)$ that the system should have provided, but that couldn't provide because future information, only available at time $(k+1)T$, was necessary to compute $e_I(k)$ using Eq.(3.13), through the hold instead of $e_d(k+1)$ in the last interval. It is important to realize that the calculated state $x(k+1)$ is EXACTLY compatible with the energy balance by construction using the result of Eq.(3.20). The only problem is the fact that the held value $e_d(k+1)$ is not equal to the value $e_I(k)$ which would have been expected by the use of discrete representation of Eq.(3.13) at the sample instant kT. On the other hand, we can set $e_d(k+2) = e_I(k)$ and use the correct value with a delay T. This operation does NOT break the passivity of the algorithm and it is possible to see that it will have little influence on the dynamics for reasonably small T since the acceleration profile will be roughly shifted by T and the inertial properties of the robotic device will behave like a low pass filter. It is then possible to calculate $e_C(k+1)$ using the value of $x(k+1)$ and then, when time $(k+2)T$ is reached, to start over the algorithm for computing the next state.

The presented strategy works fine in all situations in which the state $x(k)$ is not on a minimum of the storage function $H(\cdot)$. Problems arise instead, if at any moment of time the state $x(k)$ is indeed in a minimum. To see this, consider the simulation of a pure 1D, linear, elastic element with internal energy $H(x) = \frac{1}{2}Kx^2$ connected to a simple mass. Suppose a starting unloaded rest situation ($x(0) = 0$) and $e_d(1) = 0$. Due to the fact that $x(0)$ is in a minimum, $e_C(0) = 0$. Consider that a user would apply a motion to the robotic device in such a way that at time 1, $f_d(1)$ would result different than zero. Due to the fact that the supplied effort $e_d(1) = 0$, the exchanged power between the mass and the port-Hamiltonian controller is equal to zero and this would result again in $x(1) = 0$. The problem arises due to the fact that in this example, $e_I(0) = e_C(0) = 0$ and this implies that $e_d(2)$ will again be zero. This would result in a constant zero force applied to the robotics device which is clearly not what we wanted. This seemingly critical shortcoming of the algorithm can be nevertheless easily corrected, by modifying the choice of e_d for the next step:

$$e_d(k+2) = \frac{e_I(k) + \bar{e}_I(k)}{2} \tag{3.21}$$

where $\bar{e}_I(k)$ can be taken as the value of the port effort calculated with a first order Euler approximation of the continuous time system which we want to display in the haptic interface. In the case of the spring, this would be:

$$\bar{e}_d(k+1) = Kf_d(k+1).$$

This addition can be easily generalized for the extended structure of a general physical system. This operation would introduce some extra energy into the system but the amount of energy injected is known and it can be compensated using the techniques presented in the previous subsection.

3.4.3 Passive Coupled Behavior

From the previous considerations, it is possible to understand that at each sampling time, we have an EXACT matching between the physical energy going to the continuous time system and the virtual energy coming from the discrete time port independently of the sample time T of the discrete system. It is remarkable that the choice reported in Eq.(3.19) which is very simple and at the same time attractive due to the fact that it corresponds to position measurements, in practice gives such a powerful and at the same trivial result. This means that we can passively interconnect the two systems in such a way that *independently* of the sampling time and its relation with the characteristics of the interconnected systems, the two systems would be energetically consistent at each sampling time and no energy would be created by the sampling and hold procedure. The only energy leak is due to the fact that the discrete time system has no way whatsoever to predict the value of the continuous time system at the interconnection port and this implies that only at the end of the sample period will have an exact measure of the energy it supplied to the continuous time system. This gives rise to the problems reported in Sec. 3.4.1. These are structural problems, due to the fact that the data obtained by sampling can never give enough information in order to causally reconstruct the physical behavior of the interconnection as it would happen with an acausal Shannon reconstructor. They can be either compensated by book-keeping of the energy in excess supplied to the continuous time system or by a continuous time damping circuit which can be easily built with a couple of operational amplifiers directly connected to the power amplifiers used to drive the continuous time system.

Summarizing, using the strategy illustrated in this section it is possible to sample a continuous port-Hamiltonian system preserving passivity in its discrete counterpart. Furthermore it is possible to interconnect the discrete passive virtual environment to the continuous haptic device through a Sample & Hold algorithm that does not introduce ANY extra energy. It is therefore possible to build an haptic interface that is intrinsically passive independently of the sample period that is used to implement the virtual environment. Furthermore, exploiting the port-Hamiltonian formalism, it is possible to implement a very wide class of physical systems, both linear and non linear, as virtual environments. Notice that no extra layer has been added to the haptic interface to preserve passivity and that no limitation to the impedance that can be achieved has been introduced. Each potential source of extra energy has been simply implemented in a passive way, improving the transparency of the haptic interface.

3.4.4 Dealing with Quantization Errors

In Sec. 3.4.2 it has been shown that in order to perform a passive interconnection between a discrete and a continuous port-Hamiltonian system position measurements are needed. Position information is very often obtained by means of encoders. In this section we will show that the quantization error associated with the encoders leads to the production of some extra energy into the system and, therefore, to a loss of passivity. We will propose a strategy, introduced in [261], to interconnect continuous and discrete port-Hamiltonian systems without the production of any extra energy even in case of unreliable position information.

Encoders are affected by quantization errors that cause a position measurement error which can be modeled by an additive bounded disturbance $w(t)$. Let

$$\|w(t)\| \leq W \; \forall t$$

where W is a finite positive constant. We have that:

$$q(t) = \bar{q}(t) + w(t) \tag{3.22}$$

where $\bar{q}(t)$ and $q(t)$ represent the real value of the position and the measure respectively.

If we use the output of the encoders to calculate $f_d(kT)$ as in Eq.(3.19), we obtain:

$$\begin{aligned} f_d(k+1) &= -\frac{q(k+1) - q(k)}{T} = \bar{f}_d(k+1) - \frac{w(k+1) - w(k)}{T} = \\ &= \bar{f}_d(k+1) + \delta(k+1) \end{aligned} \tag{3.23}$$

Because of the quantization error the discrete flow in Eq.(3.23) is the sum of two terms: the ideal flow ($\bar{f}_d(k+1)$) and a spurious term ($\delta(k+1)$). Since $w(\cdot)$ is bounded, the spurious term is bounded as well and we have:

$$\|\delta(k)\| \leq \frac{2W}{T} \quad \forall k$$

Let us investigate the energetic behavior of the interconnection during a sample period $[kT, (k+1)T]$. We have that, referring to Fig. 3.14, the energy $\Delta H_c(k+1)$ flowing into $\mathcal{H}_c$ is:

$$\begin{aligned} \Delta H_c(k+1) &= H_c(k+1) - H_c(k) = \int_{kT}^{(k+1)T} e^T(\tau) f(\tau) d\tau = \\ &= -e_d^T(k+1)\bar{f}_d(k+1)T = \bar{H}_d(k) - \bar{H}_d(k+1) = -\Delta\bar{H}_d(k+1) \end{aligned} \tag{3.24}$$

On the other hand, the discrete energy increment $\Delta H_d(k+1)$ is equal to:

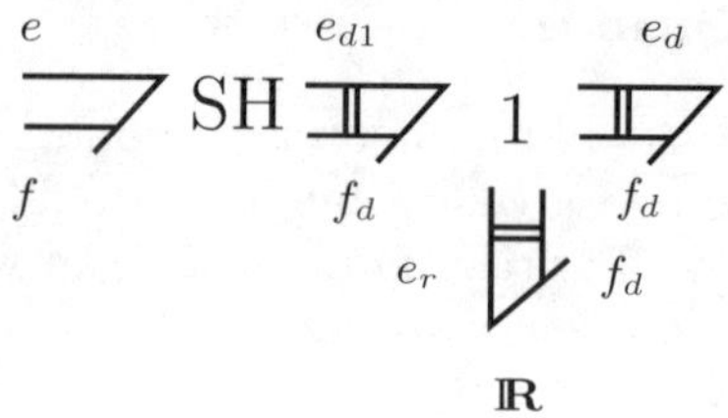

Fig. 3.15. The sample & hold plus dissipation

$$\Delta H_d(k+1) = H_d(k+1) - H_d(k) = e_d^T(k+1) f_d(k+1) T =$$

$$= e_d^T(k+1)\bar{f}_d(k+1)T + e_d^T(k+1)\delta(k+1)T = \Delta\bar{H}_d(k+1) + \Delta H_s(k+1) \tag{3.25}$$

Comparing Eq.(3.24) and Eq.(3.25), we can see that:

$$\Delta H_c(k+1) = -\Delta H_d(k+1) + \Delta H_s(k+1)$$

We have an additional term ($\Delta H_s(k)$) due to the spurious term in $f_d(k)$ and, therefore, there is no more energetic consistency between continuous and discrete domains, namely it is no more true that the amount of energy supplied to the continuous system is exactly equal to the amount of energy extracted from the discrete one. This term can lead to production of extra energy in the interconnection and, therefore, to the loss of passivity.

In order to recover passivity in the interconnection, we need to somehow dissipate the extra energy produced during the sampling. We will, therefore, consider the scheme represented in Fig. 3.15 (in a bond graph notation) where we endowed the interconnection with a dissipative element which must be properly designed. Since the spurious term $\delta(\cdot)$ is bounded, we can dissipate the maximum amount of energy introduced because of quantization error. Since:

$$\Delta H_s(k+1) = e_d^T(k+1)\delta(k+1)T$$

we have that in the worst case

$$\Delta H_{sw}(k+1) = \|e_d^T(k+1)\frac{2W}{T}T\| = \|e_d^T(k+1)2W\|$$

We can design the dissipative element such that it dissipates ΔH_{sw} at each sample period. We have that some energy can be produced, because of quantization error, in the energetic flow between the port (e, f) and the port (e_{d1}, f_d). On the other hand, the maximum amount of energy that can be produced is dissipated through the port (e_r, f_d) and, therefore, the energy flowing through the port (e_d, f_d) will be lower or equal than that flowing through the port (e, f). This means that the behavior of the interconnection between the continuous port (e, f) and the discrete port (e_d, f_d) is passive.

This approach is working fine but it is somehow over conservative and it could lead to poor performances. In fact, the quantization error on the encoders does not always cause production of energy but it can also cause dissipation. It is clear that the latter behavior does not affect the passivity of the interconnection. With the proposed approach, instead, the worst case produced energy is always dissipated, disregarding the contribution introduced by the disturbances. A less conservative scheme for the interconnection can be obtained if we have a measure of the continuous flow f; we can get this measure by means of analogic flow sensors (e.g. tachometric dynamos) or of some estimation algorithms such as a state variable filter.
By means of flow measure/estimation we can compute:

$$\Delta H_c(k+1) = \int_{kT}^{(k+1)T} e^T(\tau) f(\tau) d\tau$$

and compare it with

$$-\Delta H_d(k+1) = -e_d(k+1)^T f_d(k+1)T$$

If $\Delta H_c(k+1) > -\Delta H_d(k+1)$, then the quantization error produced some extra energy and the amount $\Delta H_d(k+1) + \Delta H_c(k+1)$ must be dissipated. If $\Delta H_c(k+1) \leq -\Delta H_d(k+1)$ the quantization error led to to some energy dissipation and, therefore, there is no need to activate the dissipative element of the interconnection. The main advantage of this approach with respect to the previous one, is that now we know exactly when there is need of dissipation and the exact amount of energy we need to dissipate and, therefore, we minimally act on the system degrading the performances as less as possible.

On the other hand, even the measure/estimation of the flow can be imprecise and therefore we will have:

$$f(t) = \bar{f}(t) + n(t)$$

where $n(t)$ represents the error on the measure/estimation and $\bar{f}_c(t)$ the real flow. We can assume that the error is bounded:

$$\|n(t)\| \leq N$$

where N is a finite positive constant. In this case we have that:

$$\int_{kT}^{(k+1)T} e^T(\tau) f(\tau) d\tau = \Delta \bar{H}_c(k+1) + e_d^T(k+1) \int_{kT}^{(k+1)T} n(t) = \tag{3.26}$$

$$= \Delta \bar{H}_c(k+1) + \Delta H_n(k+1)$$

where $\Delta \bar{H}_c(k+1)$ is the energy increment due to the real flow. We have a spurious term $\Delta H_n(k+1)$ due to the error on the measurement of the flow. This term is bounded and, similarly to $\Delta H_s(k+1)$, depends on the effort:

$$\|\Delta H_n(k+1)\| \leq \|e_d^T(k+1)NT\| := \Delta H_{nw}(k+1)$$

Because of the uncertainty introduced on the measure of $\Delta H_c(k+1)$ we can not any longer exactly predict when there has been production of energy. We can write the following algorithm for the design of the dissipative element in the interconnection:

If $\Delta H_d(k+1) + \Delta H_c(k+1) > 0$ **then**
dissipate $\Delta H_d(k+1) + \Delta H_c(k+1) + \Delta H_{nw}(k+1)$
else
dissipate $\Delta H_{nw}(k+1)$

We always need to dissipate ΔH_{nw} to be assured that there is no production of energy.

Remark 3.8. The algorithm gets less and less conservative the more the flow measure is reliable. Notice that if ΔH_{nw} is bigger than ΔH_{sw}, the proposed algorithm is over conservative with respect to the constant dissipation of ΔH_{sw}. We have to choose which algorithm to use depending on the reliability of flow measure/estimation.

3.5 Delayed Virtual Environments

It can happen that the virtual environment dynamics are very complex and that the computation time needed to run its simulation exceeds the sample period. In these cases we speak of *delayed virtual environment*. Several researchers addressed this problem and various methodologies have been proposed: in [201] an extension of the virtual coupling technique, based on input and output strict passivity has been proposed, and in [13] a wave-model based approach has been proposed.

Consider the energetic representation of an haptic interface reported in Fig. 3.8, where the virtual environment is given by a passively discretized port-Hamiltonian system which is interconnected in a passive way to the continuous haptic device using the techniques reported in Sec. 3.4. It is possible to extend this scheme in order to be able to implement an haptic interface that preserves passivity also in case of delayed virtual environments.

3.5.1 The Effect of Delayed Output

As illustrated in Sec. 2.2.3, a port-Hamiltonian system is a passive system in case there is no delay on the output. Let us consider a port-Hamiltonian represented by Eq.(3.13) where the power port by means of which the system energetically interacts with the rest of the world is represented by an effort/flow pair (e_I, f_I). Assume, furthermore, that the system has impedance causality (i.e. flow in / effort out). In case of delay we have that the output power variable at the interaction port is:

$$e_{I\delta}(t) = e_I(t - \delta)$$

The non delayed system would be passive with respect to the input/output pair (e_I, f_I). It can be easily proven that delay in the output destroys passivity of the port-Hamiltonian system. In fact:

$$\begin{aligned} P(t) &= -e_{I\delta}(t)f_I(t) = -(e_I(t) + \underbrace{e_{I\delta}(t) - e_I(t)}_{\bar{e}(t)})^T f_I(t) = \\ &= -e_I^T(t)f_I(t) - \bar{e}^T(t)f_I(t) = \frac{dH}{dt} + \underbrace{f_R(t)R(x)f_R(t) - \bar{e}^T(t)f_I(t)}_{P_{diss}} \end{aligned} \tag{3.27}$$

Specific choices of the input variable can lead to a negative value of P_{diss} and therefore to a production of extra energy and to a loss of passivity. We can therefore conclude that when there is a certain delay in the computation of the output (i.e. in case of delayed virtual environment), port-Hamiltonian systems with an interaction power port represented by an effort/flow pair cannot be safely used to implement an haptic interface because they are not passive.

Consider a power port (e, f) where e and f represent an effort and a flow respectively. As illustrated in Sec. 1.5, it is possible to define the scattering waves as:

$$s^+ = \frac{N^{-1}}{\sqrt{2}}(e + Zf)$$

$$s^- = \frac{N^{-1}}{\sqrt{2}}(e - Zf)$$

where

$$Z = NN$$

is a positive definite matrix representing the impedance of the scattering transformation.

The following power balance holds:

$$P(t) = e^T(t)f(t) = \frac{1}{2}\|s^+(t)\|^2 - \frac{1}{2}\|s^-(t)\|^2 \tag{3.28}$$

and we can interpret $s^+(t)$ as an incoming power wave and $s^-(t)$ as an outgoing power wave. A power port represents an exchange of energy between the system and the rest of the world and Eq.(3.28) shows that this exchange can be equally represented both by an effort/flow pair and by scattering variables as proven in Theorem 1.35. A definition of passivity based on scattering variables can be given. If (e, f) is the power port by means of which a system

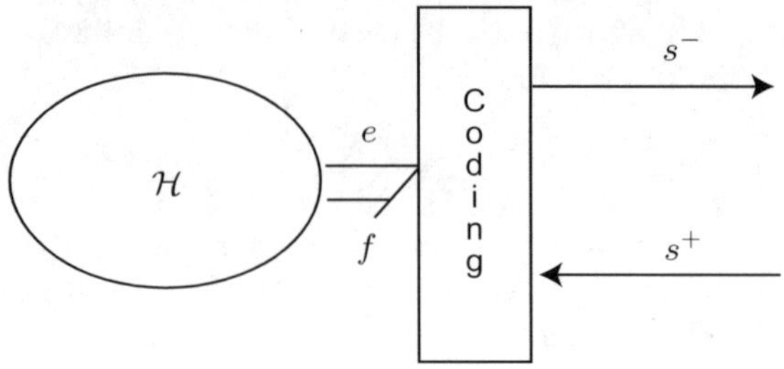

Fig. 3.16. Codification of power variables

interacts with the rest of the world and if (s^+, s^-) is its corresponding scattering representation, we have that a system is passive if and only if there exists a finite positive constant β such that:

$$\int_0^t \frac{1}{2}\|s^-(\tau)\|^2 d\tau \leq \int_0^t \frac{1}{2}\|s^+(\tau)\|^2 d\tau + \beta \tag{3.29}$$

Loosely speaking, a system is passive if and only if the outgoing energy is bounded by the incoming energy, namely if and only if there is no internal production of energy.

One could think to deal with delayed virtual environment by simply discretizing a port-Hamiltonian system with power variables as input/output and then to treat the delay in the same way as in telemanipulation (see next chapter for many more details). In Fig. 3.16 we can see how the problem can be tackled in telemanipulation. The power variables of the port of the port-Hamiltonian system are coded into scattering variables and then sent through the communication channel. The port-Hamiltonian system is passive and it has a passive behavior at the power port. By using the scattering representation of the power port and by transmitting scattering waves, this passive behavior is simply conserved during the communication, independently of any delay. Unfortunately, when dealing with delayed virtual environments, the power variables at the power port are not consistent (in the sense that they refer to different instants of time) and this causes, as shown in Eq.(3.27), a non passive behavior of the system at the port. If we used the scheme in Fig. 3.16, the scattering waves would simply replicate the non passive behavior of the systems, leaving the problem unsolved. The delay has to be treated therefore in a different way in case of delayed virtual environment. The reason of this discrepancy is that while in telemanipulation the problem concerning delay is in the transmission of power variables and, therefore, something *external* to the system, in case of delayed port-Hamiltonian systems the delay is intrinsic into the dynamics of the system, *internal* to the system, and in this case the scattering framework has to be embedded in the model of the system.

In order to deal with this internal delay we will model a port Hamiltonian system modeling the power port (e_I, f_I), by means of which it interacts with the rest of the world, by the corresponding equivalent scattering representation

(s_I^+, s_I^-). Since Eq.(3.28) holds, we can write:

$$P_I + P_C + P_R = \frac{1}{2}\|s_I^+\|^2 - \frac{1}{2}\|s_I^-\|^2 + e_C^T f_C + e_R^T f_R = 0 \tag{3.30}$$

which represents the power balance reported in Eq.(3.10) when the interaction port is represented by a pair of scattering variables. By straightforward calculation starting from Eq.(3.13) we can get:

$$\begin{pmatrix} s^- \\ f_C \end{pmatrix} = \begin{pmatrix} S_1 & S_2 \\ S_3 & S_4 \end{pmatrix} \begin{pmatrix} s^+ \\ e_C \end{pmatrix} \tag{3.31}$$

where:

$$S_1 = (BN^{-1} + N)^{-1}(BN^{-1} - N)^{-1} \tag{3.32}$$

$$S_2 = \sqrt{2}(BN^{-1} + N)^{-1}A \tag{3.33}$$

$$S_3 = \frac{CN^{-1}}{\sqrt{2}}(I - (BN^{-1} + N)^{-1}(BN^{-1} - N)) \tag{3.34}$$

$$S_4 = D - CN^{-1}(BN^{-1} + N)^{-1}A \tag{3.35}$$

and now

$$\dot{H}(x) + f_R^T R(x) f_R = -(\frac{1}{2}\|s_I^+\|^2 - \frac{1}{2}\|s_I^-\|^2) = P$$

Consider now a port-Hamiltonian system in the form of Eq.(3.31) and suppose that there is a delay in the computation of the output. We have that:

$$\begin{cases} s_{I\delta}^-(t) = 0 & t < \delta \\ s_{I\delta}^-(t) = s_I^-(t-\delta) & t \geq \delta \end{cases}$$

where we reasonably assumed that when the output is not yet available because of the delay, the virtual environment presents 0 on the output buffer.

Proposition 3.9. *Port-Hamiltonian systems whose interaction port is represented by a pair of scattering variables are passive independently of any delay* $\delta > 0$ *on the output.*

Proof. Since the non delayed system is passive we have that condition (3.29) holds and therefore that:

$$\int_0^t \frac{1}{2}\|s_I^-(\tau)\|^2 d\tau \leq \int_0^t \frac{1}{2}\|s_I^+(\tau)\|^2 d\tau + \beta \;\forall t > 0 \tag{3.36}$$

Let us now consider the delayed output, $s_{I\delta}^-(t) = s_I^-(t-\delta)$. We have that:

$$s_{I\delta}^-(t) = 0 \;\forall t \in [0, \delta]$$

We can write:

$$\int_0^t \frac{1}{2}\|s_{I\delta}^-(\tau)\|^2 d\tau \leq \int_0^t \frac{1}{2}\|s_I^-(\tau)\|^2 d\tau \leq \int_0^t \frac{1}{2}\|s_I^+(\tau)\|^2 d\tau + \beta$$

which implies:

$$\int_0^t \frac{1}{2}\|s_{I\delta}^-(\tau)\|^2 d\tau \leq \int_0^t \frac{1}{2}\|s_I^+(\tau)\|^2 d\tau + \beta$$

and therefore the delayed system is passive.

Thus, considering the scattering representation of the interaction power port, passivity is preserved even in case of delay on the output.

3.5.2 Passive Discretization of Port-Hamiltonian Systems in Scattering Representation

The aim of this subsection is to modify the passivity preserving discretization algorithm proposed in Sec. 3.4.1 in order to provide a passive discretization for port-Hamiltonian systems in scattering form.
If we rewrite Eq.(3.30) for the discrete case, we have:

$$\frac{1}{2}\|s_I^+(k)\|^2 - \frac{1}{2}\|s_I^-(k)\|^2 + e_C^T(k)f_C(k) + e_R^T(k)f_R(k) = 0 \tag{3.37}$$

Furthermore, during the interval k, we have to consider a constant state $x(k)$ corresponding to the continuous time state $x(t)$. This implies that during the interval k, the dissipated energy will be equal to $Tf_R^T(k)R(x(k))f_R(k)$ and the supplied energy will be equal to $-T(\frac{1}{2}\|s_I^+(k)\|^2 - \frac{1}{2}\|s_I^-(k)\|^2)$. In order to be consistent with the energy flows, and as a consequence conserve passivity, we need therefore a jump in internal energy $\Delta H(k)$ from instant kT to instant $(k+1)T$ such that:

$$\Delta H(k) = -Tf_R^T(k)R(x(k))f_R(k) - T(\frac{1}{2}\|s_I^+(k)\|^2 - \frac{1}{2}\|s_I^-(k)\|^2)$$

This implies that the new discrete state $x(k+1)$ should belong to an energetic level such that:

$$H(x(k+1)) = H(x(k)) + \Delta H(k)$$

We can indicate the set of possible energetically consistent states, which can be found solving the previous equation in $x(k+1)$, again as

$$I_{k+1} := \{x \in \mathcal{X} \quad s.t. \quad H(x) = H(x(k)) + \Delta H(k)\}.$$

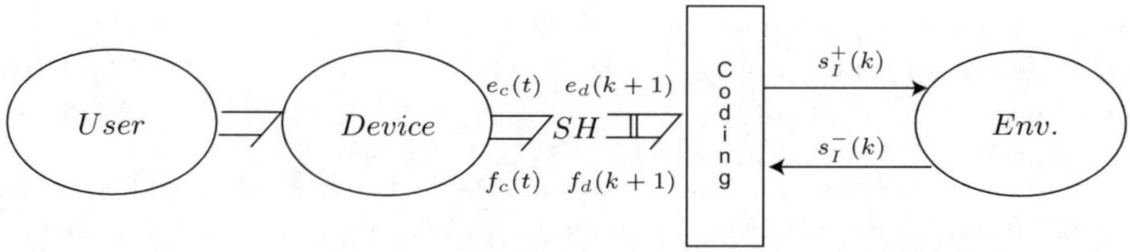

Fig. 3.17. The final scheme

Furthermore, from the discrete equivalent of Eq.(3.31), we have that:

$$f_C(k) = S_3 s_I^+(k) + S_4 e_C(k) \tag{3.38}$$

and therefore, for consistency with the continuous dynamics in which $f_C(t) = \dot{x}(t)$, the next state $x(k+1)$ should be such that:

$$f_C(k) = \lim_{T \to 0} \frac{x(k+1) - x(k)}{T} \tag{3.39}$$

where we considered the definition of the right derivative. The set I_{k+1} can be either empty or have more solutions. These two situation are treated exactly in the same way as in Sec. 3.4.1.

As a summary of the procedure just outlined, we hereafter algorithmically explain the way the discrete system can be integrated

1. Given an initial state $x(k)$, we set $e_C(k) = \frac{\partial H}{\partial x}(x_k)$.
2. Using the value of the system input $s_I^+(k)$ and the previously calculated $e_C(k)$, we can calculate $s_I^-(k)$, the output of the interaction port, and $f_C(k)$ using the discrete representation of Eq.(3.31)
3. $f_C(k)$ is then used to calculate the next state $x(k+1)$ using the procedure explained at the beginning of this subsection.

In Fig. 3.17 is represented the general scheme for an intrinsically passive port-Hamiltonian based haptic interface.

The user exchanges energy with the haptic device which is energetically coupled to the virtual environment. Since Proposition 3.9 holds, the discrete system is passive (in a discrete sense) even if the output power wave is delayed because of computational delay. The interconnection between continuous and discrete domain is made through the element SH which is the passive Sample & Hold described in Sec. 3.4.2. Since the interaction of the human operator with the virtual environment will have place through power variables (i.e. effort and flow) we endowed the scheme with a coding block which is used to interface the power variables based port with the scattering based port. In case (very frequent in haptics) that the virtual environment has an impedance causality, we have that from $f_d(k+1)$ and $s_I^-(k)$ we compute $s_I^+(k)$ and $e_d(k+1)$. Notice that now the coding procedure is safe. In fact the coding/decoding procedure is, by definition, such that the energetic behavior at the port (s_I^+, s_I^-) and the one at the port (e_d, f_d) are exactly the same (there is energetic synchronism); since Proposition 3.9 holds we have that the behavior at the port (s_I^+, s_I^-) is passive and therefore we will have a passive behavior at the port (e_d, f_d) independently of any computational delay.

The scheme provides a passive haptics scheme for delayed virtual environment without *any* extra layer to dissipate extra energy produced by the delay. The scheme is based on natural and intuitive energetic consideration and it has the advantage that transparency of the virtual environment is affected only by the unavoidable dynamical effect associated to the delay and it is *not* perturbed by any extra dynamics.

Remark 3.10. In case there is no computational delay on the output, we have that using either power variables or scattering waves leads to the same behavior of the system. The scheme in Fig. 3.17 can be, therefore, considered a generalization of the scheme proposed in Sec. 3.4.

Remark 3.11. Since the scheme proposed in Sec. 3.4 is passive independently of the sample period, one could think to avoid the delayed output problem by simply increasing the sample period. This is true but performances of the system are affected by the sample period used: the higher is the sampling rate, the more realistic will be the simulation. In order to have an efficient haptic interface, therefore, it is necessary to keep the sample period as small as possible taking into account, in the implementation, possible computational delays .

3.6 Force Scaling in Port-Hamiltonian Based Haptic Interfaces

Haptic interfaces can be successfully used for assisting the user in the execution of a given task. In several applications (e.g. surgery) it is requested that the user moves a tool along a certain path with a high degree of precision; if the trajectory to track is implemented as a virtual environment, the user can feel the path he has to track and therefore he/she can precisely follow it. *Virtual fixtures*, introduced in [248], are perceptual overlays that help the human operator to execute a task with the required degree of precision. In other words, a virtual fixture is a simulated constraint that guides the user over a preferred path enhancing its tracking performance. They have been used for increasing the precision and/or the speed in path following and in positioning tasks [21], for training [236] and for improving performances in telemanipulation tasks [248, 1].

Virtual fixtures can be interpreted as virtual environments, characterized by a certain stiffness with which the user is interacting during the execution of a task. Their role is to influence the operator's natural behavior by assisting him/her. Nevertheless, as pointed out in [220], by constraining the user's motion, a virtual fixture also limits the user's control over a given task. Thus, because of an error in the design phase, it is possible that the fixture is misplaced and that therefore it may have a negative effect on the task execution. Furthermore, it may happen that there are certain regions where the user wants to defy the virtual fixture in order either to avoid an area or to reach a position that is not on the path marked by the fixture. It is then necessary to be able to scale the force imposed by the virtual constraint on the user in order to increase the level of user's control over the task and in order to allow him/her to reach off-fixture targets. In [220] impedance fixtures are considered and the force scaling issue is addressed by changing the stiffness of the virtual constraint; several algorithms for varying the stiffness are considered and compared experimentally.

When using an intrinsically passive haptic interface any non passive operation during the execution of the task should be avoided in order to preserve a stable behavior. Thus, while a virtual fixture characterized by a certain stiffness can be passively implemented, changing the stiffness of a virtual environment is a non-passive operation as detailed in Sec. 5.5. In [220], it is reported that the users have some difficulties in using the system when they have to go back to the virtual constraint from the off fixture target, namely when the stiffness of the virtual environment they are interacting with increases; we think that this is related to the regenerative, non passive, effect associated to the stiffness increase.

In this section we replace the Sample & Hold strategy proposed in Sec. 3.4, that allows a lossless energy transfer between the haptic device and the virtual environment, with the power scaled interconnection. We will prove that in this way it is possible to scale both the velocity imposed by the user and the force transmitted by the virtual environment without affecting the passivity of the overall scheme. In fact, in this way force scaling doesn't derive from an alteration of the physical properties of the virtual environment but from the interposition of a tunable layer, whose behavior doesn't affect the passivity of the overall system, between the haptic device and the virtual environment. Using the proposed strategy it is possible to scale the feeling perceived by the user while preserving a stable behavior of the system; in particular, it is possible to scale the constraining effect of a virtual fixture in a safe, intrinsically stable way.

We will use the following notation for the discrete derivative and the discrete integral:

$$dg(k) = \frac{g(k+1) - g(k)}{T} \qquad I_h^k g = \sum_{i=h}^{k-1} g(i)T$$

where g is a generic sequence.

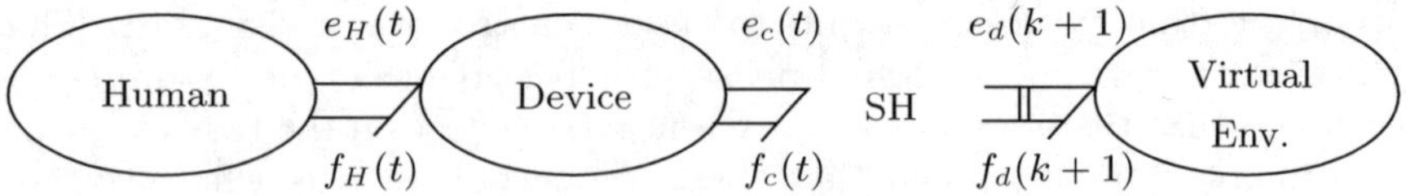

Fig. 3.18. Energetic representation of a haptic display

3.6.1 Power Scaling in Port-Hamiltonian Based Haptic Interfaces

Consider the haptic interface represented in Fig. 3.18, where the haptic device and the virtual environment are represented as a passive and a passively discretized port-Hamiltonian systems respectively which are passively interconnected using the Sample & Hold technique reported in Sec. 3.4.
Since the haptic device and the virtual environment are passive systems the following energy balances hold:

$$\int_0^{nT} (e_H^T(\tau) f_H(\tau) + e_c^T(\tau) f_c(\tau)) d\tau = H_c(x_c(nT)) - H_c(x_c(0)) + E_{diss}^c(nT) \tag{3.40}$$

and

$$I_0^n e_d^T(k+1) f_d(k+1) = H_d(x_d(nT)) - H_d(x_d(0)) + E_{diss}^d(nT) \tag{3.41}$$

where x_c and x_d are the states of the haptic device and of the virtual environment respectively, H_c and H_d are lower bounded functions representing the energy stored in the continuous haptic device and in the virtual environment and E_{diss}^c and E_{diss}^d are nonnegative functions representing the energy dissipated by the systems; we assumed that $E_{diss}^c(0) = 0$ and that $E_{diss}^d(0) = 0$, namely that at the beginning the haptic device is at rest and that the virtual environment is at equilibrium. Furthermore, when using the interconnection strategy reported in Sec. 3.4 we have that the following balance holds:

$$I_0^n e_d^T(i+1) f_d(i+1) = -\int_0^{nT} e_c^T(\tau) f_c(\tau) d\tau \qquad \forall n \in \mathbb{N} \tag{3.42}$$

which says that at any sampling instant the energy extracted by the continuous haptic device is supplied to the virtual environment and viceversa and,therefore, that no extra energy is produced in the interconnection. Scaling the interaction means to scale the exchange of energy between the haptic device and the virtual environment which, on its turn, means scaling the effort and the flow exchanged between the device and the virtual environment. Thus, we propose to use the following *power scaled Sample & Hold* strategy:

$$\begin{cases} f_d(k+1) = -\alpha \frac{q(k+1)-q(k)}{T} \\ e_c(t) = \frac{1}{\beta} hold(e_d(k+1)) \end{cases} \qquad \alpha, \beta \in \mathbb{R}^+ \tag{3.43}$$

where α and β are factors that allow to scale the interaction of the user with the environment. If we are interested in scaling only the force transmitted to the user (as in the case of virtual fixtures studied in [220]) we can set $\alpha = 1$ and tune β for achieving the desired force scaling. Using this interconnection, we have that

$$-\frac{1}{\alpha\beta} I_0^n e_d^T(i+1) f_d(i+1) = \int_0^{nT} f_c^T(\tau) e_c(\tau) d\tau \qquad \forall n \in \mathbb{N} \tag{3.44}$$

which means that the energy extracted by the virtual environment is scaled and supplied to the haptic device. Thus, the exchange of energy with the virtual environment is perceived by the user scaled by a factor $1/\alpha\beta$. Thus the interconnection strategy reported in Eq.(3.43) is not passivity preserving because, in general, the power extracted from one side can be amplified and supplied to the other side, leading to a production of energy. Nevertheless, the following result can be proven:

Proposition 3.12. *If the haptic device is a passive system, if the virtual environment is rendered as a discrete passive system and if the interconnection between the continuous and the discrete part is made through the Sample and Hold strategy reported in Eq.(3.43), then the overall haptic interface is passive.*

Proof. Since the haptic device and the virtual environment are passive systems, both Eq.(3.40) and Eq.(3.41) hold and, consequently, for each sampling instant we have that:

$$\begin{aligned} &\alpha\beta \int_0^{nT} e_H^T(\tau) f_H(\tau) d\tau + \alpha\beta \int_0^{nT} e_c^T(\tau) f_c(\tau) d\tau + I_0^n e_d^T(i+1) f_d(i+1) = \\ &= \alpha\beta[H_c(x_c(nT)) - H_c(x_c(0))] + H_d(x_d(nT)) - H_d(x_d(0)) + \\ &+ \alpha\beta E_{diss}^c(nT) + E_{diss}^d(nT) \end{aligned} \tag{3.45}$$

Since the interconnection between the haptic device and the virtual environment is made through Eq.(3.43) we have that, using Eq.(3.44):

$$I_0^n e_d^T(i+1) f_d(i+1) = -\alpha\beta \int_0^{nT} e_c^T(\tau) f_c(\tau) d\tau \tag{3.46}$$

Using Eq.(3.46) in Eq.(3.45) and taking into account that $\alpha\beta E_{diss}^c(\cdot)$ and $E_{diss}^d(\cdot)$ are nonnegative functions, it follows that:

$$\begin{aligned} &\alpha\beta \int_0^{nT} e_H^T(\tau) f_H(\tau) d\tau \geq \alpha\beta[H_c(x_c(nT)) - H_c(x_c(0))] + \\ &+ H_d(x_d(nT)) - H_d(x_d(0)) \end{aligned} \tag{3.47}$$

whence

$$\int_0^{nT} e_H^T(\tau) f_H(\tau) d\tau \geq H_c(x_c(nT)) - H_c(x_c(0)) + \frac{1}{\alpha\beta}[H_d(x_d(nT)) - H_d(x_d(0))] \tag{3.48}$$

which proves that the system is passive with respect to the storage function

$$H_c(\cdot) + \frac{1}{\alpha\beta} H_d(\cdot)$$

The scaled interconnection is NOT a passive element but, nevertheless, it can be safely used in the implementation of an haptic interface. This happens because the effect of the scaling is to "mask" the virtual environment and to make it appear to the human operator as if it acted at a different power scale transmitting, in particular, to him/her a scaled force. However, this amplification/attenuation of the power transmitted does NOT modify the kind of dynamic behavior of the virtual environment which keeps on being passive. Changing the stiffness of the virtual environment, instead, deeply influence the kind of dynamic behavior of the virtual environment and the passivity of the overall haptic interface cannot be guaranteed anymore.

3.6.2 Variable Scaling

An abrupt change in the perception of the virtual environment can disturb the user and therefore, as suggested in [220] for virtual fixtures, a gradual attenuation of the scaling has to be preferred. Thus, it is necessary to consider variable factors that gradually scale the interaction with the virtual environment. In other words, we should use the following *variable power scaled Sample & Hold* interconnection:

$$\begin{cases} f_d(k+1) = -\alpha(k) \frac{q(k+1)-q(k)}{T} \\ e_c(t) = \frac{1}{\beta(k)} hold(e_d(k+1)) \end{cases} \tag{3.49}$$

where $\alpha(k)$ and $\beta(k)$ are the bounded scaling factors and

$$\begin{aligned} \underline{\alpha} \leq \alpha(k) \leq \overline{\alpha} \quad & \underline{\alpha}, \overline{\alpha} \in \mathbb{R}^+ \\ \underline{\beta} \leq \beta(k) \leq \overline{\beta} \quad & \underline{\beta}, \overline{\beta} \in \mathbb{R}^+ \end{aligned} \tag{3.50}$$

In this case, the effect of the power scaling interconnection is no more static and, therefore, the variation of α and β can introduce some dynamic effects which could destroy the intrinsic passivity of the haptic interface. Luckily, if some not very restrictive conditions on the kind of virtual environment are satisfied, we can still preserve the passivity of the overall scheme.
The following lemma will be useful in the following:

Lemma 3.13. *Let* $f : \mathbb{R} \mapsto \mathbb{R}$ *be a real function such that*

$$\int_0^t \dot{f}(\tau)d\tau \geq -\delta \quad \forall t \quad \delta \in \mathbb{R}^+, \delta < \infty \tag{3.51}$$

and let $\gamma : \mathbb{R} \mapsto \mathbb{R}^+$ *be a scaling function such that*

$$\underline{\gamma} \leq \gamma(t) \leq \overline{\gamma} \quad \forall t \quad \underline{\gamma}, \overline{\gamma} \in \mathbb{R}^+ \tag{3.52}$$

Suppose that the function $f(\cdot)$ *has a finite number of critical points, namely that its derivative changes sign a finite number of times. Then*

$$\int_0^t \gamma(\tau)\dot{f}(\tau)d\tau \tag{3.53}$$

is lower bounded.

Proof. Let $sign(\cdot)$ indicate the sign function. Consider an interval $[t_0, t_1]$ such that $sign(\dot{f}(t)) = const.$ $\forall t \in [t_0, t_1]$. We can distinguish three cases. If $sign(\dot{f}(t)) = 1$ then

$$\begin{aligned}\int_{t_0}^{t_1} \gamma(\tau)\dot{f}(\tau)d\tau &\geq \underline{\gamma}(f(t_1) - f(t_0)) = \underline{\gamma}(f(t_1) - f(0)) + \underline{\gamma}(f(0) - f(t_0)) \geq \\ &\geq -\underline{\gamma}\delta + \underline{\gamma}(f(0) - f(t_0))\end{aligned} \tag{3.54}$$

If $sign(\dot{f}(t)) = 0$ then

$$\int_{t_0}^{t_1} \gamma(\tau)\dot{f}(\tau)d\tau = 0 \tag{3.55}$$

If $sign(\dot{f}(t)) = -1$ then

$$\int_{t_0}^{t_1} \gamma(\tau)\dot{f}(\tau)d\tau \geq \overline{\gamma}(f(t_1) - f(0)) + \overline{\gamma}(f(0) - f(t_0)) \geq -\overline{\gamma}\delta + \overline{\gamma}(f(0) - f(t_0)) \tag{3.56}$$

Every interval $[0, t]$ can be split up in the following way

$$[0, t] = [0, t_1] \bigcup \cup_{i=1}^{p} I_{p_i} \bigcup \cup_{j=1}^{n} I_{n_j} \bigcup \cup_{k=1}^{z} I_{z_k} \tag{3.57}$$

where

$$\begin{aligned} I_{p_i} &= [t_i, \bar{t}_i] \quad \bar{t}_i > t_i \quad sign(\dot{f}(t)) = 1 \forall t \in I_{p_i} \\ I_{n_j} &= [t_j, \bar{t}_j] \quad \bar{t}_j > t_j \quad sign(\dot{f}(t)) = -1 \forall t \in I_{n_j} \\ I_{z_k} &= [t_k, \bar{t}_k] \quad \bar{t}_k > t_k \quad sign(\dot{f}(t)) = 0 \forall t \in I_{n_k} \end{aligned} \tag{3.58}$$

and where $sign(\dot{f}(t)) = const.\ \forall t \in [0, t_1]$. Let Γ be defined as:

$$\Gamma = \begin{cases} \overline{\gamma} \text{ if } sign(\dot{f}(t)) = -1 \forall t \in [0, t_1] \\ 0 \text{ if } sign(\dot{f}(t)) = 0 \forall t \in [0, t_1] \\ \underline{\gamma} \text{ if } sign(\dot{f}(t)) = 1 \forall t \in [0, t_1] \end{cases} \tag{3.59}$$

Thus we can write

$$\begin{aligned} \int_0^t \gamma(\tau)\dot{f}(\tau)d\tau \geq \Gamma \int_0^{t_1} \dot{f}(\tau)d\tau + \sum_{i=0}^{p} \int_{t_i}^{\bar{t}_i} \gamma(\tau)\dot{f}(\tau)d\tau + \\ + \textstyle\sum_{j=1}^{n} \displaystyle\int_{t_j}^{\bar{t}_j} \gamma(\tau)\dot{f}(\tau)d\tau \geq -\Gamma\delta + \sum_{i=0}^{p}(-\underline{\gamma}\delta + \underline{\gamma}(f(0) - f(t_i))) + \\ + \textstyle\sum_{j=0}^{n}(-\overline{\gamma}\delta + \overline{\gamma}(f(0) - f(t_j))) = -\Gamma\delta - p\underline{\gamma}\delta - n\overline{\gamma}\delta + \\ + \underline{\gamma} \textstyle\sum_{i=0}^{p}(f(0) - f(t_i)) + \overline{\gamma} \sum_{j=0}^{n}(f(0) - f(t_j)) \end{aligned} \tag{3.60}$$

$f(t_i)$ and $f(t_j)$ are finite since they are critical points. Furthermore, because of the hypothesis, $p < \infty$ and $n < \infty$ and therefore the sums are finite and consequently the integral is lower bounded.

Thanks to Lemma 3.13 it is possible to build interfaces where the haptic flow can be variably scaled. In fact we can prove the following result:

Proposition 3.14. *Consider the scheme reported in Fig. 3.8 and suppose that the Sample & Hold strategy is implemented through the strategy reported in Eq. (3.49). Suppose that the energy function that characterizes the virtual environment has a finite number of critical points. Then the overall haptic interface is passive.*

Proof. Set $\alpha(k)\beta(k) = \gamma(k)$, $\underline{\alpha\beta} = \underline{\gamma}$ and $\bar{\alpha}\bar{\beta} = \overline{\gamma}$. We have that

$$\gamma(i) \int_{iT}^{(i+1)T} e_c^T(\tau) f_c(\tau) d\tau = -e_d^T(i+1) f_d(i+1)T \tag{3.61}$$

Furthermore we have that

$$\begin{aligned} \int_{iT}^{(i+1)T} \gamma(i)e_H^T(\tau)f_H(\tau)d\tau + \int_{iT}^{(i+1)T} \gamma(i)e_c^T(\tau)f_c(\tau)d\tau + e_d^T(i+1)f_d(i+1)T = \\ = \gamma(i)[\int_{iT}^{(i+1)T} e_H^T(\tau)f_H(\tau)d\tau + \int_{iT}^{(i+1)T} e_c^T(\tau)f_c(\tau)d\tau] + e_d^T(i+1)f_d(i+1)T = \\ \gamma(i)(\Delta H_c(i)) + \Delta H_d(i) + \gamma(i)E_{diss}^c(i) + E_{diss}^d(i) \end{aligned} \tag{3.62}$$

where

$$\Delta H_c(i) = H_c(x_c((i+1)T)) - H_c(x_c(iT))$$

and

$$\Delta H_d(i) = H_d(x_d((i+1)T)) - H_d(x_d(iT))$$

Using Eq.(3.61) in Eq.(3.62) we have that:

$$\int_{iT}^{(i+1)T} e_H^T(\tau) f_H(\tau) d\tau \geq \Delta H_c(i) + \frac{1}{\gamma(i)}(\Delta H_d(i)) \tag{3.63}$$

Summing over $i = 0, 1, \ldots, n-1$ we have that

$$\int_0^{nT} e_H^T(\tau) f_H(\tau) d\tau \geq H_c(x_c(nT)) - H_c(x_c(0)) + \underbrace{\sum_{i=0}^{n-1} \frac{1}{\gamma(i)} \Delta H_d(i)}_{\Delta \bar{H}_d} = \tag{3.64}$$

$$= H_c(x_c(nT)) - H_c(x_c(0)) + \bar{H}_d(nT) - \bar{H}_d(0)$$

H_c is a lower bounded function and, since $H_d(\cdot)$ has a finite number of critical points and since $\gamma(i)$ is a bounded positive function, a straightforward application of the discrete version of Lemma 3.13 allows to conclude that also $\bar{H}_d(\cdot)$ is lower bounded. Thus, the energy balance reported in Eq.(3.64) proves that the overall system is passive.

Remark 3.15. The assumptions made by the lemma are not very restrictive and they are satisfied by the majority of energy functions associated to the virtual environments that are simulated in haptic interfaces. Nevertheless, they are necessary. In fact, we can passively simulate a physical environment characterized by an energy function $H(x) = \sin(x)$ but in case of variable scaling, this virtual environment would lead to an unstable interaction.

If the conditions reported in Proposition 3.14 are satisfied, it is possible to use the interconnection reported in Eq.(3.49) for changing the perception of the virtual environment at each instant while preserving passivity, and consequently a stable behavior, of the haptic interface.

3.7 Simulations

In this section some simulations validating the results obtained in Sec. 3.4, Sec. 3.5 and Sec. 3.6 are proposed.

The first set of simulations is intended to validate the results obtained in Sec. 3.4.2 about the algorithm proposed to achieve a passive interconnection between continuous and discrete time domains and the algorithm proposed in Sec. 3.4.4 in order to deal with possible quantization error on the encoders. We connected a simple mass with an initial state $p_0 = 1\ Kgm/sec$ connected

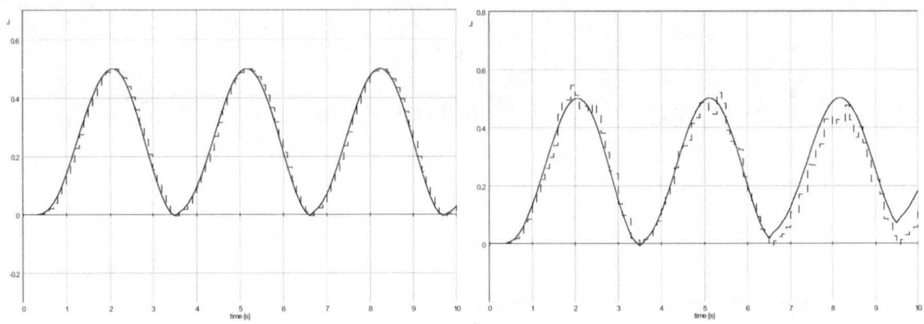

(a) Ideal energetic behavior

(b) Energetic behavior with quantization noise

Fig. 3.19. Energetic behavior in the ideal and in the noisy case. Continuous Energy (solid) and Discrete Energy (dashed)

to a discrete spring, obtained by discretizing a continuous spring by means of the algorithm proposed in Sec. 3.4.1; the sample period is $T = 0.1$ *sec.* In Fig. 3.19 we can see the energetic behavior at the interconnection both in case of ideal encoder and when there is a quantization error with maximum amplitude of 0.05. In picture (a) we can notice that, using the Sample & Hold strategy proposed in Sec. 3.4.2, continuous and discrete energy exactly match at the sampling time and, thus, a lossless transfer of energy between continuous and discrete time domains is achieved. In picture (b) we can notice that the quantization on the encoder introduces a mismatch between continuous and discrete energy, as reported in Sec. 3.4.4, and the discrete energy can be greater that the continuous one meaning that the quantization leads to production of extra energy when going from the continuous to the discrete domain, leading thus to a loss of passivity in the interconnection. In Fig. 3.20 we can see the energetic behaviors in case compensated interconnection is activated. Picture (a) shows the energetic behavior in case the worst case produced energy is dissipated. We can see that discrete energy is always lower than continuous one and, therefore, passivity has been recovered despite of quantization noise on the encoders. Nonetheless, the energetic behavior is quite different from the ideal one. In picture (b) we can see the energetic behavior in the compensated interconnection in case the information of flow sensors (or estimation algorithms) is unreliable (we simulated an unreliability with maximum amplitude of 0.01). We can see that the behavior is passive and closer to the ideal case than the one we have in case we dissipate the worst case produced energy.

The next set of simulations shows that the behavior of a coupled continuous/discrete system is passive (and, therefore, stable) according to the

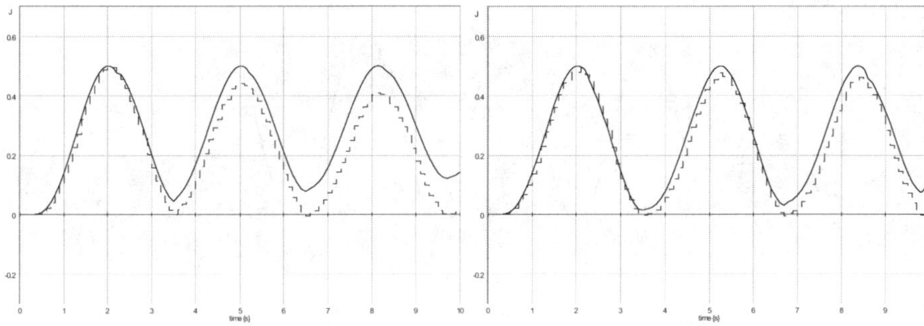

(a) Worst Case produced energy dissipation

(b) Using Unreliable flow sensors

Fig. 3.20. Compensations. Continuous (solid) and discrete (dashed) energy

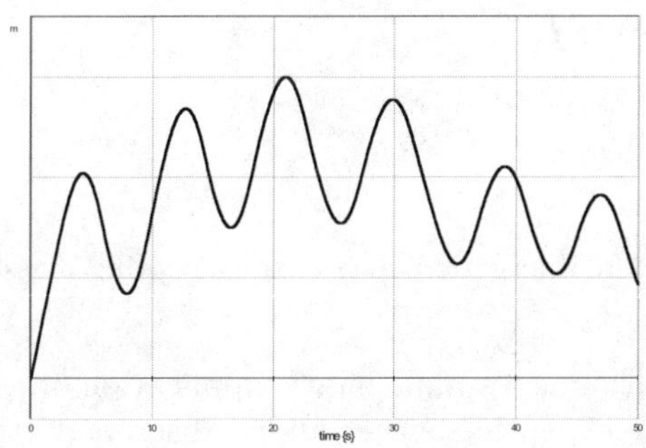

Fig. 3.21. Position of the mass

results obtained in Sec. 3.4.3. We first simulated a mass spring system where the mass is a continuous system and the spring is implemented as a discrete port-Hamiltonian system obtained as described in Sec. 3.4.1. The mass and the spring are connected as illustrated in Sec. 3.4.2 and the energy booking procedure shown in Sec. 3.4.1 is used. The mass has an initial state $p_0 = 1\ Kgm/sec$ and, therefore, it oscillates around the equilibrium point of the discrete spring. In a first simulation the sample time is set to $T = 0.5\ sec$ and we can see in Fig. 3.21 the position of the mass. We can notice that the behavior of the system is quite different from the one we would have if we used a continuous spring. In the next simulation we set the sample time at $T = 0.1\ sec$ and we can see from Fig. 3.22 that in this case the behavior of the system is much more similar to the one of its continuous counterpart. Decreasing the sample time, the behavior of the system gets closer and closer to the one of its continuous counterpart but we can *always* guarantee the stability of the overall system disregarding the sample time.

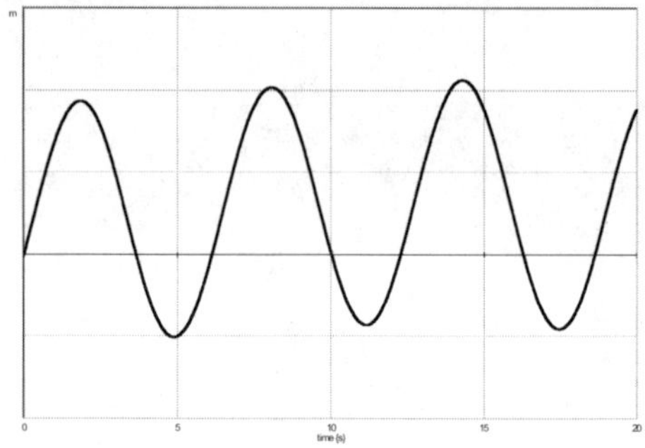

Fig. 3.22. Position of the mass

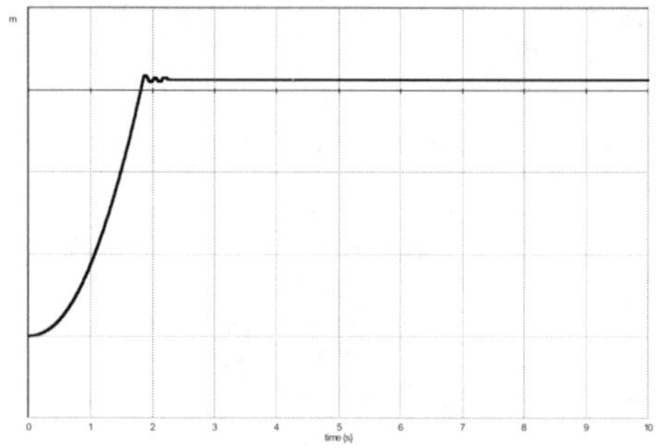

Fig. 3.23. Energetic behavior of the spring-mass system

The next simulation is a simple haptic application of our scheme. We have a mass, a continuous haptic device, and a discrete virtual environment, a virtual wall. We implemented the virtual wall as a discrete port-Hamiltonian made up of a parallel of a very stiff spring ($k = 100000\ N/m$) and of a damper ($b = 30\ Nsec/m$). The simulation is one-dimensional: the wall is at the position 0 and the haptic device is at an initial position $x_0 < 0$ and is pushed by a constant force towards the wall; the sample frequency is $f = 25\ Hz$. We can see from Fig. 3.23 that the position of the haptic device increases until it meets the wall; when the haptic device gets in touch with the virtual wall it stops when the force applied by the virtual wall balances the force applied to the system by the human operator. A stable behavior is achieved even if the sampling frequency is quite low.

The following set of simulations aims to validate the results obtained in Sec. 3.5 on the effect of delayed output on passivity and on the use of scattering variables to achieve a passive behavior independently of the output delay. Consider a simple mass, with an initial state $p_0 = 1\ mKg/sec$, connected with a discrete spring. The spring is implemented with the strategy depicted in Sec. 3.4.1 and the interconnection between continuous and discrete time is made by the passive sample and hold algorithm proposed in Sec. 3.4.2. The sample period is $T = 10\ ms$. In Fig. 3.24 we introduced a computational delay

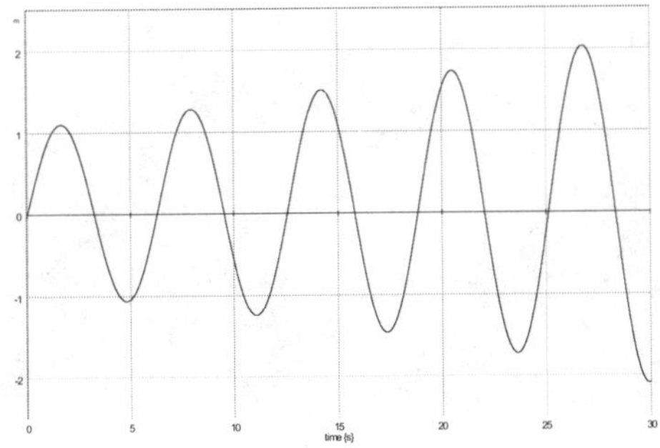

Fig. 3.24. Position of the mass in case of a computational delay $\delta = 0.05$ on the output of the spring

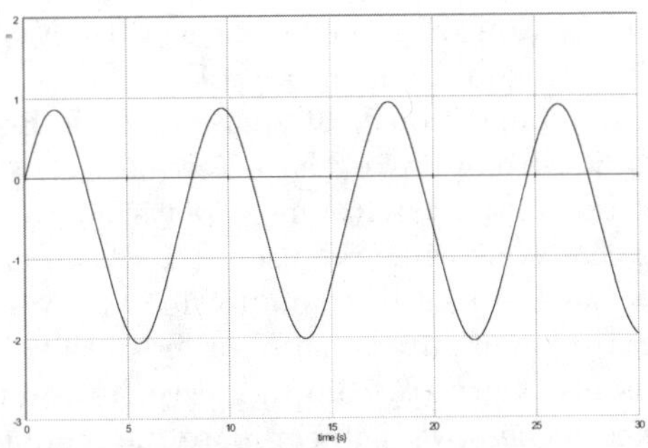

Fig. 3.25. Position of the mass in case of computational delay $\delta = 0.05sec$: scattering representation

$\delta = 0.05$ *sec* and we can see that passivity is lost and an unstable behavior is introduced. In the next simulation we represented the discrete spring by the scattering formalism and we can see that the computational delay doesn't change passivity properties and that a passive, and therefore stable, behavior is preserved. The simulation results are reported in Fig. 3.25. The shift with respect to the normal behavior of a mass-spring system is due to the dynamical effect deriving from the output delay.

The next simulation implements a virtual wall. The operator pushes the haptic device (a mass) with a constant force and, at $x = 0.3$ m the system meets a virtual wall (implemented with a very stiff spring and a high damper). We can see in Fig. 3.26 that, even if the operator keeps on pushing, the mass stands still at $x = 0.3$ m because of the action of the virtual wall. The virtual wall is simulated with $T = 0.01$ *sec* and there is a computational delay of 3 sample periods. The behavior of the overall system is passive despite of the delayed output of the virtual wall.

The last set of simulations aims at validating the results obtained in Sec. 3.6. We consider a 1-DOF intrinsically passive haptic interface. The haptic device is a simple mass of 1 Kg characterized by some friction that is

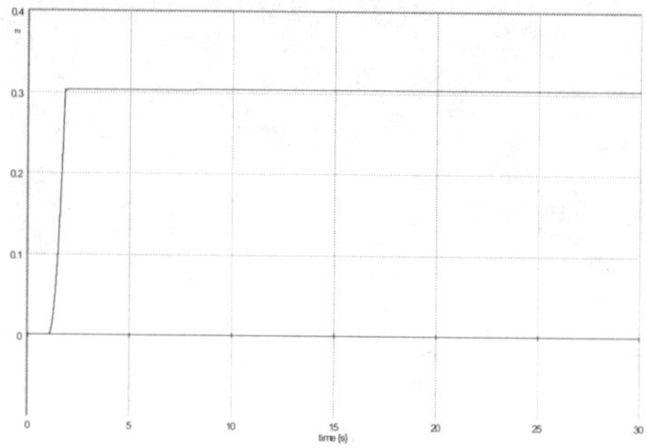

Fig. 3.26. Virtual Wall

modeled as a viscous friction with coefficient $b = 0.1\ Ns/m$. The virtual environment is the discrete equivalent of a spring that is joined to the haptic device through the interconnection strategy reported in Eq.(3.49). The overall system is passive and, therefore, it is characterized by a stable behavior.

We suppose that the virtual environment plays the role of a very simple 1-DOF virtual fixture that constrains the user at $x = 0\ m$ and that the stiffness of the virtual fixture has been set to a small value ($10\ N/m$) to allow the user to easily reach off-fixture configurations. In order to take back the user to the fixture position, it is necessary to scale the force transmitted to the user. In order to show its effectiveness, we will compare the scaling strategy proposed in Sec. 3.6 with the effect that would be obtained by directly changing the stiffness of the virtual environment. Thus, suppose that the the user is keeping the haptic device at the position $x_0 = 1\ m$ applying the relatively small force $F = 10\ N$. At time $t = 10\ s$, the user has to be taken back to the fixture position ($x = 0\ m$) and, therefore, the force applied from the virtual environment has to be scaled.
In order to test the effect of the change of stiffness, we don't introduce any scaling between the haptic device and the virtual environment and, therefore, we use the lossless interconnection described in Sec. 3.4.
In Fig. 3.27 the effect of an abrupt change of stiffness is shown. The stiffness of the virtual environment is taken instantaneously from a value of $10\ N/m$ to a value of $400\ N/m$; this corresponds to scaling the force perceived by the user of a factor 40. Nevertheless, as explained in Sec. 5.5, this operation corresponds to an introduction of energy into the system and, therefore, to a loss of passivity and this leads to an unstable behavior. The loss of passivity due to the change of stiffness is not related to the way in which the change takes place. In fact, we can see in Fig. 3.28, where the stiffness is changed linearly from $10\ N/m$ to $400\ N/m$, that the overall behavior is still unstable. The instability is related to the variation of the stiffness, which introduces energy into the system and which consequently breaks the passivity, and not to the way in which it is changed.

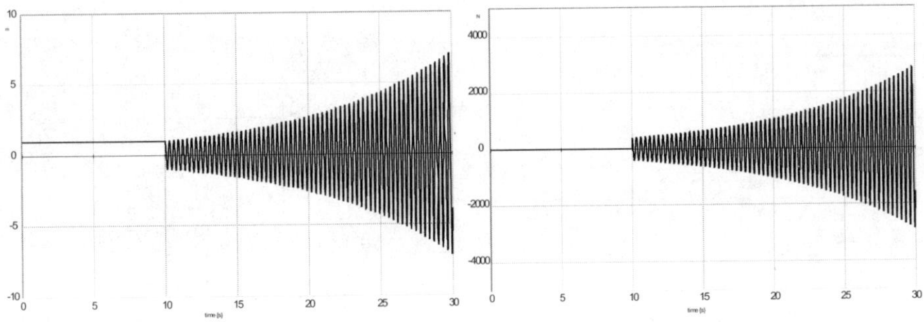

(a) Position

(b) The force exchanged with the user

Fig. 3.27. An abrupt change of stiffness

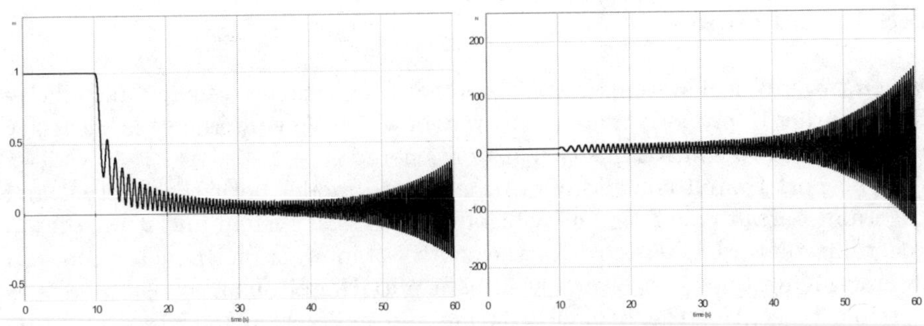

(a) Position

(b) The force exchanged with the user

Fig. 3.28. A gradual change of stiffness

In the next simulation we don't change the stiffness of the virtual fixture but we scale the force transmitted to the human using the proposed strategy. Since we are interested in scaling only the force, we set $\alpha = 1$ and we let β changing linearly from the value 1 to the value 1000. The results of the simulation are shown in Fig. 3.29.

We can see that, even if the force scaling factor is much bigger than that implemented with the stiffness variation, the behavior of the overall system is stable. This happens because the force scaling is achieved without violating the passivity and the conditions of Proposition 3.14 are satisfied.

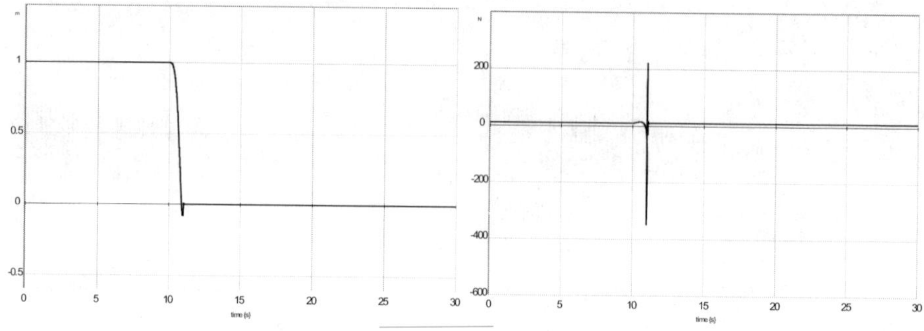

(a) Position

(b) The force exchanged with the user

Fig. 3.29. The proposed scaling strategy

3.8 Conclusions

When two physical systems get into contact they exchange energy and, therefore, in order to properly manage the way in which they interact it is necessary to model and to control this exchange of energy.

The port-Hamiltonian formalism allows to model both the internal and the input-output energetic behavior of any physical system and it provides a general framework where to treat the control of interaction. In fact, a physical interacting systems can be modeled as a port-Hamiltonian system and it is possible to control the behavior during the interaction by interconnecting to the plant a port-Hamiltonian controller whose role is to properly shape the energetic properties of the system. Thus, thanks to the port-Hamiltonian framework, it is possible to impose any kind of interacting behavior, both linear and non linear, to any kind of physical interacting system, despite of its complexity. Furthermore, it is possible to describe the controller as a virtual physical system "attached" to the plant and this gives a clear interpretation of the kind of interacting behavior imposed to the plant. A very interesting application of control of interacting systems from a port-Hamiltonian point of view can be found in [76, 75] where the problem of controlling a walking robot is addressed.

The port-Hamiltonian formalism allows to deal, in very general terms, with the control of interaction with virtual environments, namely with the development of haptic interfaces. Thanks to the port-Hamiltonian structure and to an energetic analysis, it is possible to implement all the energy leaks of an haptic interface in a passive way and, therefore, it is possible to achieve an intrinsically passive, and therefore, characterized by a stable behavior, haptic interface without inserting any spurious dynamic effect due to passivating controllers and without limiting the impedance that can be perceived by the user. Furthermore, due to their generality, port-Hamiltonian based haptic interfaces allow to consider any kind of haptic device and to simulate any kind of virtual environment and not only linear ones. Furthermore, using the scattering formalism and power scaled interconnections, it is possible to passively deal with the problems of delayed virtual environments and of the scaling of the feeling transmitted to the user. In the PO/PC method introduced in [115] the energy leak due to the hold device was not taken into account and some assumptions on the sampling rate had to be made in order to ensure passivity. Using the concepts introduced in Sec. 3.4.2, Ryu and Hannaford improved the PO/PC strategy considering the energy that could be produced by the hold device [254].

4

Port-Hamiltonian Based Bilateral Telemanipulation

4.1 Introduction

Telemanipulation is one of the first fields of application of robotics (see [326] for an early history) and still one of the most challenging. In teleoperation a human operator has to perform a certain task on a remote environment. The human operator commands a local robotic interface (called *master*). The motion of the master is transmitted through a communication channel to a remote robot (called *slave*) which should replicate the motion of the master and perform a desired task on the remote environment. It is possible to improve performances providing to the human operator some real-time information about the interaction of the slave with the remote environment. This feedback information can be achieved in several ways (e.g. through visual displays, [146]) but the best way to improve the operator's ability is to feedback the contact force between the slave and the environment to the master side. When the force at the slave side is reflected back to the human operator, it is said that the telemanipulation is controlled bilaterally, or, more simply, that we have a *bilateral telemanipulation system*. When teleoperation is performed over a great distance, such as in undersea or in space applications, or over packet switching network, such as the Internet, the communication delay associated to the transmission of information from master side to slave side and vice-versa becomes non negligible and it can therefore destabilize the whole system. There are two main problems in the implementation of a bilateral telemanipulation system. Firstly, when performing long distance teleoperation or when using packets switching networks as Internet for the transmission of information, the non negligible time delay in the control loop can destabilize the whole system. Secondly, in most useful application the slave has to interact with a non structured remote environment (e.g. spatial exploration) and therefore it is necessary to control master and slave in such a way to guarantee that a stable behavior is achieved both in case of interaction and non interaction with the environment.

A lot of work has been done in the field of telemanipulation and several control schemes can be found in the literature. The first work dealing with time delay in telemanipulation is [90] but, since force feedback was not used, no stability problems due to transmission delay were present. In [91] the *force reflection scheme* was proposed and force feedback was used for the first time in presence of non negligible time delay. The scheme is very intuitive: the master transmits to the slave the position information and the slave feeds back to the master the force information. The instability due to the time delay was evident: delays of the order of 0.1 *sec* were shown to destabilize the overall system. In [145] the *position error scheme* is proposed: master and slave exchange position information along the communication channel and the forces applied to the robots depend on the difference between the exchanged positions. Also in this case even a small communication delay destabilizes the overall scheme. In [273] some predictive control strategies are used. In particular, the well known Smith predictor is used at the master side to predict the slave response and a simple PD controller is used at the slave side. Nonetheless, in order to use the Smith predictor, a perfect knowledge of the delay is needed and instability can arise in case the amount of time delay is not perfectly known; moreover, the fact that the slave has to interact with the environment introduces some further limitations in the scheme. In [155] the four-channel architecture was proposed. This is a very general telemanipulation scheme in which both force and velocity information is exchanged between master and slave. This scheme is very good since the slave perfectly tracks the master and the contact force between the slave and the environment is perfectly reflected at the master side in case there is no delay in the communication. On the other hand, when some delay is present in the transmission of information between master and slave sides, performances of the system dramatically decrease and instability can arise. In [235] some sliding mode techniques were adopted to deal with uncertainties and with the non negligible time delay. The scheme works very fine in case the slave is not interacting with the environment: a good tracking is achieved and the scheme is stable independently of the time delay. On the other hand, when the slave interacts with the remote environment instability can arise. In [343] both master and slave are driven by an adaptive motion/force controller and position and force setpoints are exchanged. The performances of the scheme are very good in case there is no communication delay. When a non negligible delay is present, the system is still stable in case there is no interaction between the slave and the remote environment but, in case of interaction, the telemanipulation scheme can turn to instability.

A bilateral telemanipulation system is basically a robotic system for interaction where, in addition to the usual problems related to control of interaction, there is another problem related to the presence of a communication delay between master and slave sides. Passivity theory has been exploited to tackle the destabilizing effect of the delay. In [7] scattering theory has been used to achieve a lossless communication channel independently of *any constant* transmission delay. In [215] this strategy has been further developed, the concept of wave variable has been introduced and the problem of wave reflection arising in scattering based communication channels has been addressed. Several control schemes based on passivity and on wave variables have been proposed in the literature; in [210] wave variables are coupled with predictive control in order to improve performances while preserving a passive behavior in presence of communication delay and in [52] a passivity based control scheme to guarantee position tracking is developed. In [249] the PO/PC strategy is applied to bilateral telemanipulation and in [309, 310, 311] a construction, based on wave variables, to extend the perception that can be achieved through a telemanipulation system is presented. For a deeper overview of the literature and a comparison of the various telemanipulation schemes see, for instance, [124, 10, 197].

In Chap. 3 it has been shown that passivity theory and port-Hamiltonian formalism can be profitably used for the control of interacting systems. In this chapter it will be illustrated how to use the geometric scattering introduced in Sec. 1.5 and the port-Hamiltonian based control of interaction in order to achieve an telemanipulation scheme which is intrinsically passive independently of any communication delay and of the interaction with any passive, possibly unknown, remote environment [299]. Port-Hamiltonian based telemanipulation allows to describe a very huge class of telerobotic systems where master and slave can be any kind of mechanical systems, as robotic hands [271]. Furthermore it is shown how to use the techniques reported in Chap. 3 in order to consider digital controllers and discrete communication channels like the Internet. A discrete scattering based communication channel is defined and a communication strategy such that passivity is preserved independently of communication delays and packets losses is developed [263, 298]. Eventually, a passivity preserving interpolation algorithm to rebuild lost packets proposed in [261] is illustrated.

4.2 Port-Hamiltonian Based Bilateral Telemanipulation

4.2.1 An Energetic Analysis of a Bilateral Telemanipulation System

A bilateral telemanipulation system is a robotic system for interaction. It consists of a robotic interface that allows a human operator to interact, namely to exchange energy, with the environment. A schematic representation of a bilateral telemanipulation system is given in Fig. 4.1. As usual, the half arrows represent energy exchange in bond-graph formalism [237, 141]. The human operator interacts, through a robotic interface, the master, with a, possibly unknown, remote environment and the interaction is regulated by the teleoperator. In Chap. 3 it has been shown that a stable control of interaction with any passive environment can be achieved if the controller is an intrinsically passive system and it is connected in a power preserving way to the

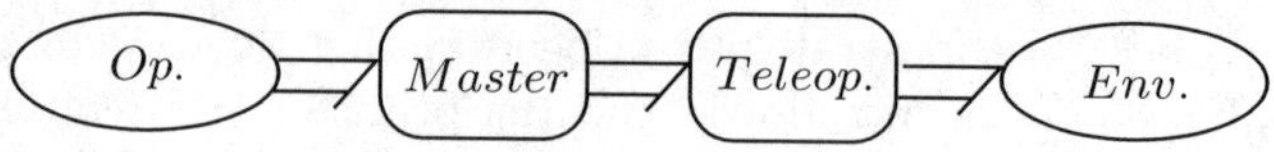

Fig. 4.1. A schematic representation of a telemanipulation scheme

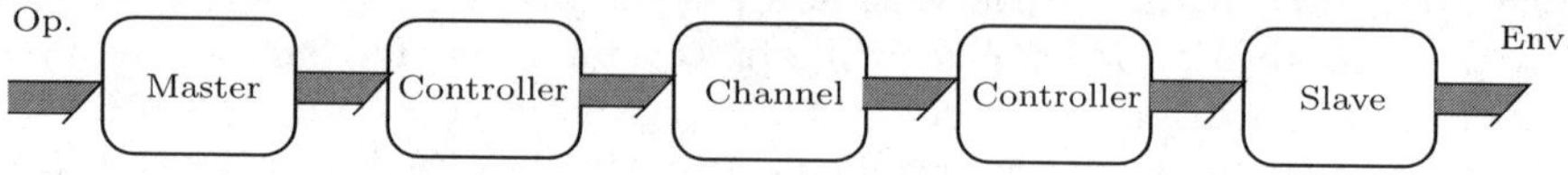

Fig. 4.2. A detailed representation of a generic bilateral telemanipulation scheme

local robotic interface. Furthermore, it has been shown that port-Hamiltonian formalism allows to take into account also possible nonlinearities of the robotic interface very easily and provides a unique framework for modeling and controlling interacting systems. Since intrinsic passivity guarantees a stable behavior both in case of free motion and interaction with unknown environments, we would like to build a telemanipulation scheme that enjoys this appealing property and therefore we would like to apply the same techniques used in Chap. 3 to the implementation of a bilateral telerobotic scheme. Since the human operator can be considered as a passive system during interacting tasks [123] and since the master device is a passive mechanical system, it is sufficient to implement the teleoperator as an intrinsically passive system and to connect it in a power preserving way (e.g. standard negative feedback, see Sec. 2.3.2) to the master robot.

The teleoperator can be decomposed in four main subsystems: two controllers, a communication channel and a robot. A more detailed representation of a generic bilateral telemanipulation scheme is given in Fig. 4.2.

In bilateral telemanipulation systems the interaction between the master and the remote environment takes place through the communication channel. In energetic terms, the master does not exchange energy directly with the environment but energy reaches the remote side through the communication channel. When the energy arrives to the remote side, it is delivered to the slave, which uses it to perform the desired task and which directly interacts with the remote environment. Thus the master receives the energy by the human operator and it exchanges energy with the communication channel while the slave receives the energy by the communication channel and it interacts with the remote environment. It is then necessary to control both the interaction between the master and the communication channel and between the slave and the remote environment and, therefore, it is necessary to endow master and slave sides with controllers that regulate the interactions of the robots. The human operator interacts with the remote environment by exchanging energy with it through all the subsystems in Fig. 4.2. The feeling perceived by the user can be deteriorated in several ways (see the next chapter for more details) the most critical of which is the rise of unstable behaviors of the telemanipulation system. Instability can be avoided by implementing an intrinsically passive scheme. Since master and slave are passive mechanical systems, in order to implement an intrinsically passive telemanipulation system, it is sufficient to passively implement the controllers for the robotic interfaces and the communication channel through which energy is exchanged and to interconnect in a power preserving way all the components.

4.2.2 Passive Control of Interaction

In Sec. 3.2 it has been shown a very effective control architecture to achieve an intrinsically passive control of interaction: the IPC. This control strategy can be successfully adopted in the implementation of a bilateral telemanipulation scheme and, therefore, the local controllers for master and slave will be implemented as IPCs and the control structure both at master and slave side is the one represented in Fig. 3.1. At the master side the plant is the robot and the human operator plays the role of the supervisory system since it injects the energy required to perform a certain task while the communication channel plays the role of the environment since the IPC controls the exchange of energy between the master and the transmission line. At the slave side, instead, the plant is the robot while the supervisory system is the communication channel, since the slave side receives the energy necessary to perform a certain task from the master side through the communication channel. The remote environment plays the role of the environment since the IPC controls the interaction, i.e. the energy exchange, between the robot and the remote environment. Summarizing, by using the IPC as local controllers it is possible to guarantee a stable behavior in the interaction between the master and the communication channel and between the slave and the remote environment. The port-Hamiltonian framework can be exploited to model master and slave sides. In fact, the robots are mechanical systems that can be modeled as port-Hamiltonian systems and the IPCs, as shown in Sec. 3.2, can be interpreted as virtual physical systems and, therefore, they can be modeled as port-Hamiltonian systems interconnected in a power preserving way to the robots that they control. Thus, since Proposition 2.29 holds, both master and slave sides can be represented as port-Hamiltonian systems and, consequently, they are passive and they can be described by:

$$\begin{cases} \dot{x} = (J(x) - R(x))\frac{\partial H}{\partial x} + g(x)u \\ y = g^T(x)\frac{\partial H}{\partial x} \end{cases} \tag{4.1}$$

where

$$u = \begin{pmatrix} u_{ch} \\ u_I \end{pmatrix} \qquad y = \begin{pmatrix} y_{ch} \\ y_I \end{pmatrix}$$

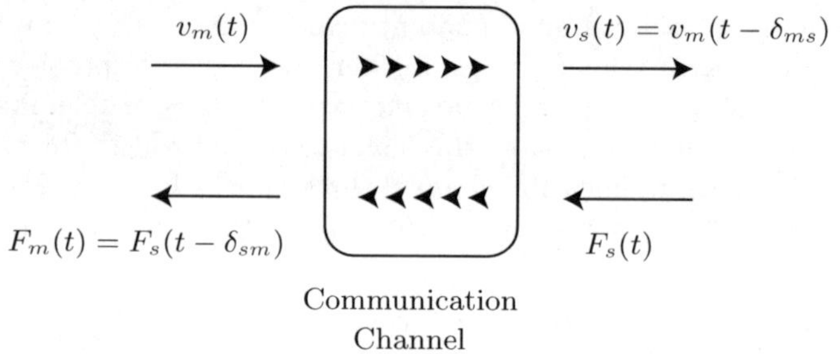

Fig. 4.3. A possible implementation of a communication channel for force reflection

the power ports (u_{ch}, y_{ch}) and (u_I, y_I) are used to exchange energy with the communication channel and either with the human (for the master side) or the environment (for the slave side) respectively.

Remark 4.1. We model master and slave sides as explicit port-Hamiltonian systems because this is sufficient for considering the most commonly used telemanipulation setups and because this allows to keep the computations reasonably simple. Nevertheless all the results provided in this chapter can be straightforwardly generalized to the case where master and slave sides needs to be modeled as implicit port-Hamiltonian systems.

4.2.3 Passive Communication Channel

Since in bilateral telemanipulation the slave has to track the master motion and the master has to feel the contact force between the slave and the remote environment, one of the first communication strategy used was to transmit velocity information from the master to the slave and force information from the slave to the master. This kind of communication channel is represented in Fig. 4.3; δ_{ms} and δ_{sm} represent the constant communication delays in the transmission from master to slave and from slave to master respectively, $v_m(t)$ and $v_s(t)$ represent the velocity at the master and at the slave side respectively and $F_m(t)$ and $F_s(t)$ represent the force at the master and at the slave side respectively. Unfortunately when using this kind of transmission line, even some small delay in the communication between master and slave can destabilize the system [91]. This destabilizing behavior can be attributed to the communication channel and it is not dependent on the particular kind of controller used to command master and slave. Making a passivity analysis of this kind of communication strategy, it is possible to understand whence instability comes. We will slightly generalize the communication strategy reported in Fig. 4.3 and we will assume that the master and slave exchange m dimensional power variables: the master transmits to the slave a flow while the slave transmits back to the master an effort. Thus:

$$\begin{aligned} f_s(t) &= f_m(t-\delta_{ms}) \\ e_m(t) &= e_s(t-\delta_{sm}) \end{aligned} \tag{4.2}$$

where δ_{ms} and δ_{sm} are the delays relative to the communication between master and slave and the other way around. Suppose that the communication channel connects two dissipative, and therefore strictly passive, elements that are modeled by

$$e_i = Bf_i \qquad i = m, s$$

where B is a $m \times m$ positive definite matrix.

Proposition 4.2. *The communication strategy reported in Eq.(4.2) leads to a non passive communication channel.*

Proof. The power flowing in the communication channel is given by:

$$P = e_m^T(t)f_m(t) - e_s^T(t)f_s(t)$$

By simple algebraic manipulations it follows that:

$$\begin{aligned} P = &\frac{1}{2}e_m^T(t)e_m(t) + \frac{1}{2}f_m^T(t)B^TBf_m(t) + e_m^T(t)(I-B)f_m(t)- \\ &-\frac{1}{2}(e_m(t)-Bf_m(t))^T(e_m(t)-Bf_m(t)) + \frac{1}{2}e_s^T(t)e_s(t) + \frac{1}{2}f_s^T(t)B^TBf_s(t)+ \\ &+e_s^T(t)(B-I)f_s(t) - \frac{1}{2}(e_s(t)+Bf_s(t))^T(e_s(t)+Bf_s(t)) \end{aligned} \tag{4.3}$$

where I represent the $m \times m$ identity matrix. Since Eq.(4.2) holds, Eq.(4.3) can be rewritten as:

$$\begin{aligned} P = &e_m^T(t)e_m(t) + f_s^T(t)B^TBf_s(t) + e_m^T(t)(I-B)f_m(t) + e_s^T(t)(B-I)f_s(t)- \\ &-\frac{1}{2}(e_m(t)-Bf_m(t))^T(e_m(t)-Bf_m(t)) - \frac{1}{2}(e_s(t)+Bf_s(t))^T(e_s(t)+Bf_s(t))+ \\ &+\frac{d}{dt}\left[\int_{t-\delta_{ms}}^{t}\frac{1}{2}f_m^T(\tau)B^TBf_m(\tau)d\tau + \int_{t-\delta_{sm}}^{t}\frac{1}{2}e_s^T(\tau)e_s(\tau)d\tau\right] \end{aligned} \tag{4.4}$$

Defining the lower bounded stored energy function H and the power dissipation P_{diss} as:

$$H = \frac{d}{dt}\left[\int_{t-\delta_{ms}}^{t}\frac{1}{2}f_m^T(\tau)B^TBf_m(\tau)d\tau + \int_{t-\delta_{sm}}^{t}\frac{1}{2}e_s^T(\tau)e_s(\tau)d\tau\right]$$

and

$$P_{diss} = e_m^T(t)e_m(t) + f_s^T(t)B^TBf_s(t) + e_m^T(t)(I-B)f_m(t) + e_s^T(t)(B-I)f_s(t) -$$

$$-\frac{1}{2}(e_m(t) - Bf_m(t))^T(e_m(t) - Bf_m(t)) - \frac{1}{2}(e_s(t) + Bf_s(t))^T(e_s(t) + Bf_s(t))$$

the following relation follows:

$$P = \frac{dH}{dt} + P_{diss} \tag{4.5}$$

which expresses the energetic behavior of the system. Recalling Def. 2.11, a system is passive if and only if the incoming power P is either stored or dissipated. However, specific choices of the input variables $f_m(t)$ and $e_s(t)$ can lead to negative values of P_{diss} and this means that the communication channel can produce energy and that therefore it is not passive

Thus the communication channel implemented using the strategy reported in Eq.(4.2) leads to an active and, therefore, potentially destabilizing subsystem. Even if the master and the slave side are passive, the communication channel can introduce, in presence of non negligible communication delays, extra energy into the system and destabilize it. On the other hand, it is still possible to use this kind of channel in bilateral telemanipulation. In fact, it is sufficient to insert a sufficient amount of damping at master and slave side in order to be able to dissipate the energy that can be introduced by the transmission line, as reported in [215]. However, this is not a good solution since it leads to over damped systems which can achieve very limited velocities and very poor performances.

In Sec. 4.2.1 it has been shown that a bilateral telemanipulation system is nothing else than a robotic interface that allows the human operator to interact with a remote environment. Furthermore, in Chap. 3 it has been explained that interaction in nothing else than exchange of energy. The aim of the communication channel is to transfer energy from the master side to the slave side. A logic consequence of a delay in the transfer of energy is that the interaction between the human operator and the remote environment will be delayed. However, it is not clear why a delay in the delivery of energy should cause production of energy and an unstable behavior. *Why, then, the communication strategy reported in Eq.(4.2) leads to a non passive communication channel from a physical point of view?*

In the communication channel represented in Fig. 4.3 each side transmits to the other side a power variable. The power delivered at the master side, therefore, is equal to the product of the power variables at the slave side. Assume, for simplicity, that $\delta_{ms} = \delta_{sm} = \delta$. Since the master transmits energy to the slave, we expect that the power at time t at the slave side is equal to the power injected into the communication channel at time $t-\delta$ at the master side, namely:

$$e_m^T(t-\delta)f_m(t-\delta) = e_s^T(t)f_s(t) \tag{4.6}$$

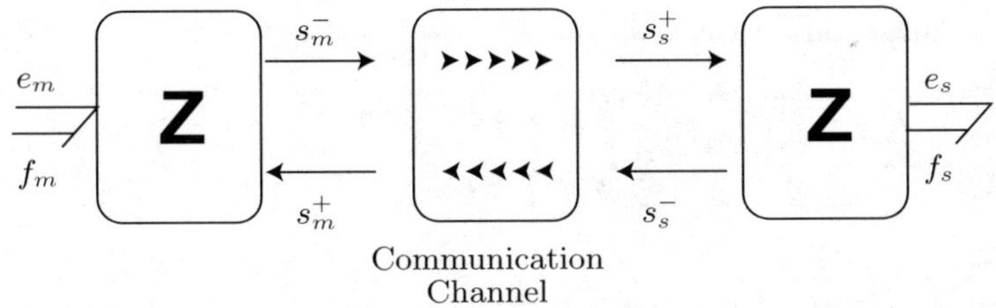

Fig. 4.4. The scattering based communication channel

but, since Eq.(4.2) holds

$$e_s^T(t)f_s(t) = e_m^T(t+\delta)f_m(t-\delta) \tag{4.7}$$

Thus, transmitting power variables, leads to the product of efforts and flows referred to different time instants and therefore to a loss of power consistency which can cause, as formally proven in Proposition 4.2, production of extra energy.

It is possible to use scattering theory to implement a passive scattering based communication channel . It has been proven in Theorem 1.35 that the power flow can be split into two scattering variables that represent an incoming and an outgoing power wave. The main idea behind the implementation of a scattering based communication channel is to split the power flows at master and slave sides into scattering variables and to transmit through the communication channel these power waves instead of power variables. In this way, the transmission line is use to transfer *directly* power and not power variables that have to be processed in order to get power. The scattering based communication channel is represented in Fig. 4.4. where $Z = NN$ is the symmetric positive defined characteristic impedance of the scattering transformation and it is used to get scattering variables from power conjugated variables. The following relations hold:

$$\begin{cases} s_m^+(t) = \frac{N^{-1}}{\sqrt{2}}(e_m(t) + Zf_m(t)) \\ s_m^-(t) = \frac{N^{-1}}{\sqrt{2}}(e_m(t) - Zf_m(t)) \end{cases} \begin{cases} s_s^+(t) = \frac{N^{-1}}{\sqrt{2}}(e_s(t) + Zf_s(t)) \\ s_s^-(t) = \frac{N^{-1}}{\sqrt{2}}(e_s(t) - Zf_s(t)) \end{cases} \tag{4.8}$$

and

$$e_m^T(t)f_m(t) = \frac{1}{2}(s_m^+(t))^T s_m^+(t) - \frac{1}{2}(s_m^-(t))^T s_m^-(t) = \frac{1}{2}\|s_m^+(t)\|^2 - \frac{1}{2}\|s_m^-(t)\|^2 \tag{4.9a}$$

$$e_s^T(t)f_s(t) = \frac{1}{2}(s_s^+(t))^T s_s^+(t) - \frac{1}{2}(s_s^-(t))^T s_s^-(t) = \frac{1}{2}\|s_s^+(t)\|^2 - \frac{1}{2}\|s_s^-(t)\|^2 \tag{4.9b}$$

The communication strategy becomes:

$$
\begin{aligned}
s_s^+(t) &= s_m^-(t-\delta_{ms}) \\
s_m^+(t) &= s_s^-(t-\delta_{sm})
\end{aligned}
\tag{4.10}
$$

It is possible to prove the following:

Proposition 4.3. *The scattering based communication channel is lossless independently of any constant communication delay.*

Proof. There are two power waves injecting energy into the communication channel and two power waves extracting energy from the communication channel; therefore, the power flowing into the communication channel is given by:

$$
P = \frac{1}{2}\|s_m^-(t)\|^2 + \frac{1}{2}\|s_s^-(t)\|^2 - \frac{1}{2}\|s_m^+(t)\|^2 - \frac{1}{2}\|s_s^+(t)\|^2 \tag{4.11}
$$

but, since Eq.(4.10) holds, Eq.(4.11) can be rewritten as:

$$
P = \frac{1}{2}\|s_m^-(t)\|^2 - \frac{1}{2}\|s_m^-(t-\delta_{ms})\|^2 + \frac{1}{2}\|s_s^-(t)\|^2 - \frac{1}{2}\|s_s^-(t-\delta_{sm})\|^2 \tag{4.12}
$$

Defining as energy of the transmission line:

$$
H = \int_{t-\delta_{ms}}^{t} \frac{1}{2}\|s_m^-(\tau)\|^2 d\tau + \int_{t-\delta_{sm}}^{t} \frac{1}{2}\|s_s^-(\tau)\|^2 d\tau
$$

it follows that:

$$
P = \frac{dH}{dt} \tag{4.13}
$$

which means that the power flowing through the communication channel is stored, namely that the system is lossless.

Thus, using the scattering variables as transmission variables, it is possible to implement the transmission as a lossless, and therefore passive, system. The losslessness of the scattering based communication channel proves the intuitive fact that a certain delay in transferring energy from one side to the other does not lead to production of energy. The power is simply transferred from one side to the other; before reaching the target side, the energy associated to the scattering variables is in the communication channel, which stores it before releasing it.

Summarizing, the transfer of energetic information from one side to the other can be done safely if and only if energy is transmitted directly, as in the case of scattering based communication channels. Power variables or other variables that yield power after that have been processed, cannot be safely transmitted because the communication delay destroys the relation between the transmitted power and the received power, as shown in Eq.(4.6) and Eq.(4.7).

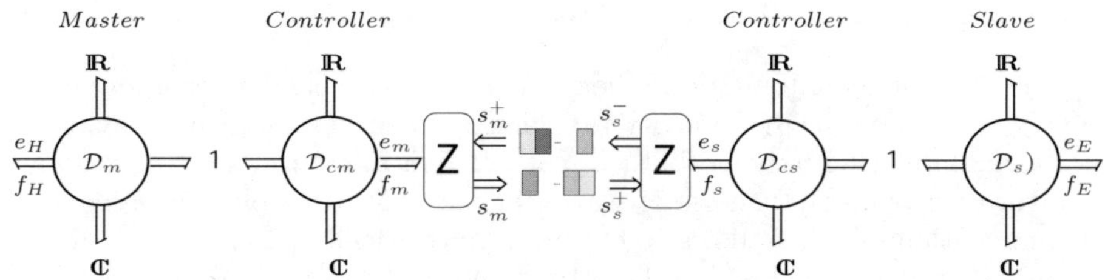

Fig. 4.5. The port-Hamiltonian Telemanipulation Scheme

4.2.4 The Intrinsically Passive Telemanipulation Scheme

Now, all the ingredients to build an intrinsically passive bilateral telemanipulation system have been introduced. The scheme is represented in Fig. 4.5 in a bond-graph notation where $\mathcal{D}_i$, $\mathbb{C}$ and $\mathbb{R}$ represent the interconnection structure, the energy storage and the energy dissipation of a port-Hamiltonian system.

Both master and slave are modeled as port-Hamiltonian systems that are interconnected in a power preserving way to intrinsically passive port-Hamiltonian impedance controllers (IPC) that allow to regulate the interactive behavior of the robots. Thus, both master and slave sides can be modeled as passive port-Hamiltonian systems. Furthermore, information is exchanged through a the lossless scattering based communication channel described in the previous section. There are two ways to connect the controllers to the communication channel in a power preserving manner:

1. Computing the effort e and s^- as a function of the flow f and of the incoming power wave s^+
2. Computing the flow f and s^- as a function of the effort e and of the incoming power wave s^+

We choose the second option and, therefore, we use IPCs whose power port interconnected to the communication channel has an impedance causality (flow in, effort out). In this way we can use the controllers proposed in Sec. 3.3 when master and slave are anthropomorphic robots or robotic hands. Nevertheless, all the considerations developed in the following can be generalized to the case in which controllers are interconnected to the communication channel using the first option.

Using the first option, the interconnection between the IPC and the communication channel consists of computing the input flow f and the outgoing power wave s^- using the output effort e and the incoming power wave s^+. Thus, at each side, the input of the IPC and the power wave to transmit are obtained by the following relations:

$$f(t) = \sqrt{2}N^{-1}s^+(t) - e(t) \tag{4.14a}$$

$$s^-(t) = \sqrt{2}N^{-1}e - s^+(t) \tag{4.14b}$$

The whole telemanipulation scheme is passive independently of any constant transmission delay. Passivity is sufficient to ensure a stable behavior of the telemanipulation scheme both in case of free motion and of interaction with the environment. The use of scattering variables to implement a lossless communication channel allows to obtain an intrinsically passive bilateral telemanipulation scheme but, on the other hand, gives rise to another problem that has to be handled: wave reflection. In physics, waves are reflected at junctions and terminations, that is at points where the impedance of the wave carrier changes. Since scattering variables represent power waves, this problem arises when using scattering based communication channels in telemanipulation. The wave reflection does not affect passivity of the overall scheme since it does not lead to production of energy but simply to a "bouncing" of the energy between master and slave side. On the other hand, this phenomenon corrupts the useful information flow and leads to an oscillatory behavior that can greatly decrease the performances of the overall system. It is therefore necessary to address the problem of matching the impedance along the overall system; this can be done either by a proper choice of the parameters of the controllers or adding additional elements.

In Sec. 4.2.2 it has been shown that master and slave sides can be represented as port-Hamiltonian systems. The IPCs are chosen with an impedance causality (flow in, effort out) at the port connected to the communication channel and, therefore, master and slave sides can be represented by a port-Hamiltonian system of the form:

$$\begin{cases} \dot{x} = [J(x) - R(x)]\frac{\partial H}{\partial x} + g(x)f \\ e = g^T(x)\frac{\partial H}{\partial x} \end{cases} \tag{4.15}$$

where f and e are the power variables representing the power exchange between the controller and the communication channel and where, in order to keep the notation simple, we have omitted to indicate the power port relative to the interaction with the environment (or with the human) since it doesn't play any role in the wave reflection phenomenon. From the scattering transformation reported in Eq.(4.8) it is possible to express power conjugated variables as a function of scattering variables:

$$\begin{cases} e = \frac{N}{\sqrt{2}}(s^+ + s^-) \\ f = \frac{N^{-1}}{\sqrt{2}}(s^+ - s^-) \end{cases} \tag{4.16}$$

Thus, replacing Eq.(4.16) in Eq.(4.15), it is possible to obtain the port-Hamiltonian model of either master or slave side in terms of scattering variables:

$$\begin{cases} \dot{x} = [J(x) - R(x)]\frac{\partial H}{\partial x} + \frac{1}{\sqrt{2}}g(x)N^{-1}(s^+ - s^-) \\ \frac{1}{\sqrt{2}}N(s^+ + s^-) = g^T(x)\frac{\partial H}{\partial x} \end{cases} \tag{4.17}$$

By simple calculations, Eq.(4.17) can be rewritten as:

$$\begin{cases} \dot{x} = [J(x) - R(x) - g(x)N^{-1}N^{-1}g^T(x)]\frac{\partial H}{\partial x} + \sqrt{2}N^{-1}g(x)s^+ \\ s^- = \sqrt{2}N^{-1}g^T(x)\frac{\partial H}{\partial x} - s^+ \end{cases} \tag{4.18}$$

It can be directly seen from Eq.(4.18) that the incoming power wave s^+ is directly fed through the output s^-; this means that the incoming power wave is reflected back on the communication channel. In order to eliminate the problem of wave reflection, it is necessary to match the impedances of the physical media crossed by the scattering waves. To this aim, we modify the IPC and to add a feedthrough term on the output of the controlled system. In this way, the robots controlled by this new IPC can be represented by the following extended port-Hamiltonian system:

$$\begin{cases} \dot{x} = [J(x) - R(x)]\frac{\partial H}{\partial x} + g(x)f \\ e = g^T(x)\frac{\partial H}{\partial x} + B(x)f \end{cases} \tag{4.19}$$

where $B(x)$ is a positive semidefinite matrix that represents a newly added dissipative term. In this case, using Eq.(4.16) and Eq.(4.19) it follows that:

$$s^- = (\frac{N}{\sqrt{2}} + B(x)\frac{N^{-1}}{\sqrt{2}})^{-1}g^T(x)\frac{\partial H}{\partial x} + (N + B(x)N^{-1})^{-1}(B(x)N^{-1} - N)s^+) \tag{4.20}$$

In order to avoid the wave reflection, it is sufficient to set:

$$B(x) = Z \tag{4.21}$$

that is, the newly added feedthrough term has to be set equal to the impedance of the communication channel. In this way, Eq.(4.20) becomes:

$$s^- = \frac{N^{-1}}{\sqrt{2}}g^T(x)\frac{\partial H}{\partial x} \tag{4.22}$$

In this way, there is no power reflection along the communication line since the impedances of master and slave side match the impedance of the scattering based transmission line.

It is possible to give a very intuitive physical interpretation of the impedance matching procedure. The feedthrough term added in Eq.(4.19) is a dissipative element that dissipates all the power that the system would try to reflect. If $B(x) = Z$ then all the power that the system tries to reflect is

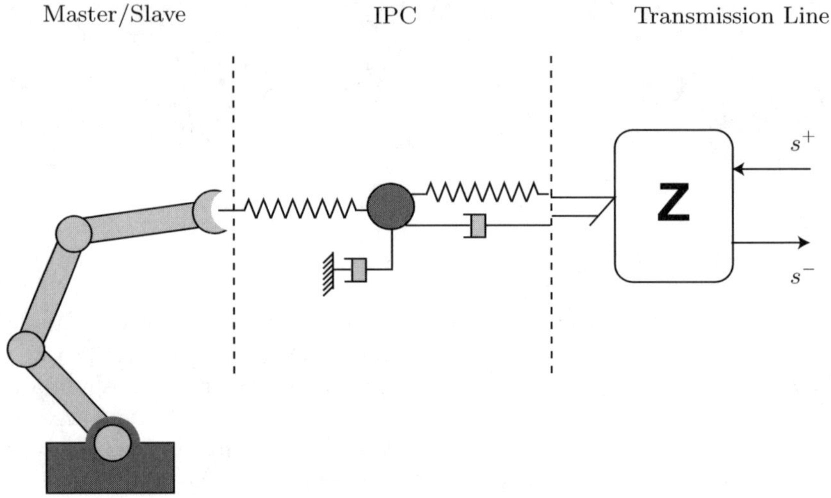

Fig. 4.6. Teleoperator with anthropomorphic robotic arms

absorbed, while if $B(x) \neq Z$, only a part of the power that the system tries to reflect is absorbed. This intuition appears even clearer, if some specific IPC is considered. In bilateral telemanipulation schemes, it happens quite often that master and slave robots are anthropomorphic robotic arms and, therefore, the IPC structure proposed in Sec. 3.3.1 can be fruitfully adopted. The impedance matching problem can be solved by adding a damper between the virtual object and the transmission line, as shown in Fig. 4.6. The spatial springs allow to assign to the interaction a desired compliance and the virtual object allows to inject some damping into the system without the need of velocity measurements. Finally, the damper inserted between the virtual object (usually called *line damper*) and the communication line has to be tuned in order to get perfect impedance matching.

In order to check the validity of the intrinsically passive port-Hamiltonian based bilateral telemanipulation scheme, some simulations are shown. Consider a one-dimensional teleoperator. Each robot is a 1 DOF system, a simple mass, and it is controlled by means of the IPC proposed in Sec. 3.3.1. The communication channel is implemented by using the scattering variables, as illustrated in Sec. 4.2.3. The communication delay is equal to 1 *sec* in both senses of transmission. In the first simulation, the human operator applies an impulsive force to the master which, because of the damping injected through the IPC, stops at a certain position. In Fig. 4.7 the behavior of the systems is shown. In figure (a) the positions of master and slave in case no line damper has been added to the IPC are shown. The behavior of the system is stable even if the communication delay is quite big and the slave reaches, after a certain delay, the position of the master. However, the performances of the system are quite bad since both master and slave are characterized by an

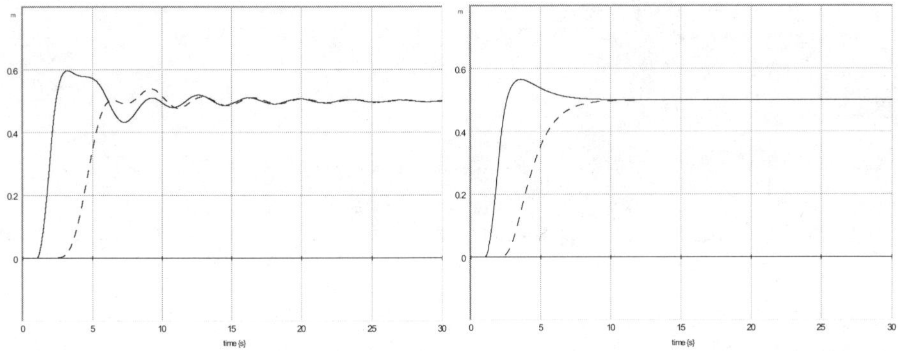

(a) No impedance matching (b) Impedance matching

Fig. 4.7. Free motion with and without impedance matching. Positions of the master (solid) and of the slave (dashed)

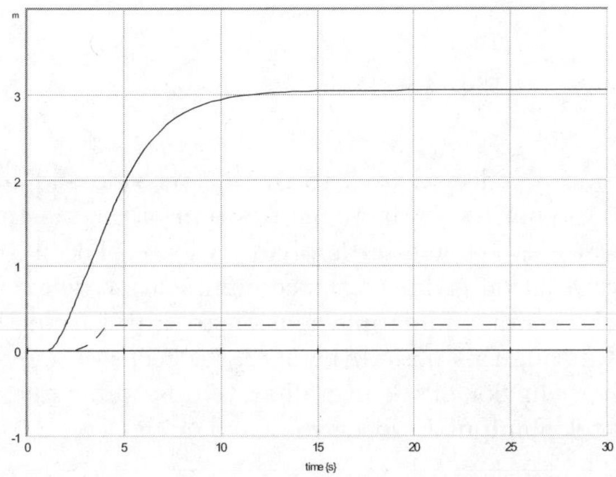

Fig. 4.8. Interaction task. Positions of the master (solid) and of the slave (dashed)

oscillatory behavior; this is due to the wave reflection phenomenon associated to the scattering based communication channel. In figure (*b*) the behavior of the overall system is shown in case the line damper has been added to the controllers. In this case the oscillations disappear, and performances are much better, since wave reflection has been removed . The next simulation is implements an interaction task. The master is pushed with a constant force and the slave interacts with a wall (implemented as the parallel of a stiff spring and of a damper) posed at the position $x = 0.3$; positions of master and slave are shown in Fig. 4.8. The slave stops when it meets the wall and the interaction takes place in a stable way because the robot is controlled by the IPC. The

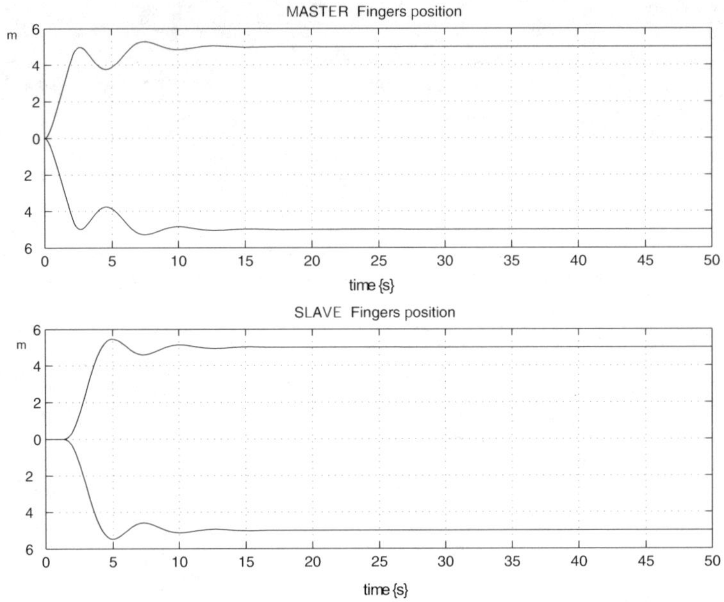

Fig. 4.9. Position of the fingers

force of interaction is reflected back to the master side and compensates the force applied to the master. In fact, the position of the master gets constant even if the operator is keeping on applying a force. The mismatch between master and slave position is due to the communication delay. In fact, it takes some time for the contact force information to travel from the slave to the master side and during this period the operator keeps on moving the master. Nevertheless, the behavior of the overall system is stable due to the intrinsic passivity of the telemanipulation scheme.

4.3 Complex Telemanipulation Systems: Telegrasping

The intrinsically passive telemanipulation scheme illustrated in the previous section works very well in systems where master and slave are anthropomorphic robot. However the scheme is much more general and can be applied also to complex telemanipulation systems. The aim of this section is to illustrate the application of the concepts reported in Sec. 4.2 to a complex telerobotic system: a telegrasping system. In this kind of telemanipulation systems, master and slave are grippers; the slave gripper has to follow the behavior of the master gripper and it has to reflect to the master the contact force due to the grasping of an object. In Sec. 3.3.2, it has been show a possible IPC structure for the control of interaction of multi fingered robotic hands. In the controller

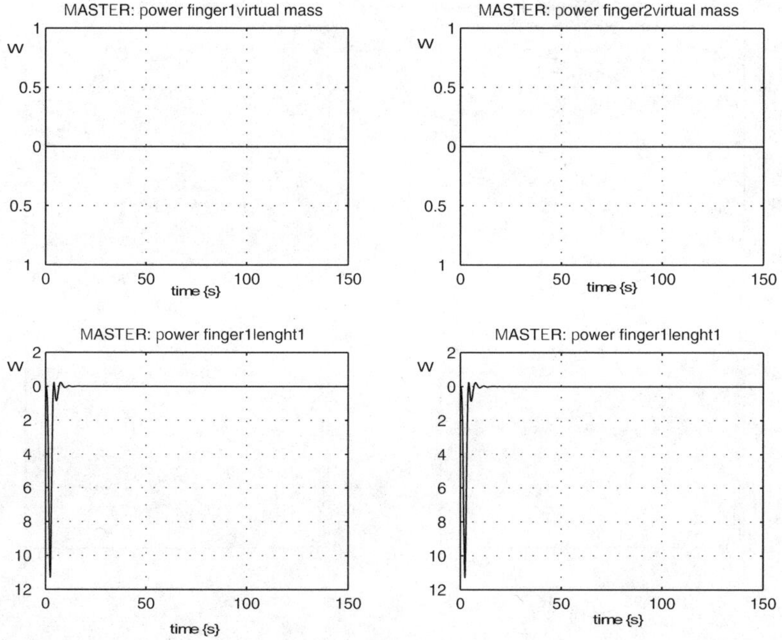

Fig. 4.10. The power at master side

proposed, illustrated in Fig. 3.7, there are $n + 1$ power ports: one of them is the relative to the virtual point and the others are the power ports we can use to vary the rest length of the variable springs that connect the n fingers to the virtual object. The length ports of the variable springs are used to vary the relative position of the fingers and the virtual port is used in order to vary the position of the whole gripper.

The main idea in the implementation of the telegrasping system is to transmit not only the power relative to the port of the virtual point but also that associated to the ports of the springs rest length. In such a way all the actions made on the master gripper, namely a variation of the relative position of the fingers or a variation of the position of the whole system, can be transmitted to the slave side and each force applied to the fingers of the slave gripper, can be reflected to the master side. Thus, in order to implement a telegrasping system, $n + 1$ scattering based communication channels have to be used: one for the virtual point, similarly to what has been done in the previous section, and n for the ports associated to the rest length of the spatial springs that connect the fingertips to the virtual object. The problem of wave reflection arising with the use of scattering based communication channels has to be solved and this is done as shown in Sec. 4.2, namely adding feedthrough dissipative terms that match the impedances of the scattering based transmission lines. Consider a simple gripper with two fingers controlled

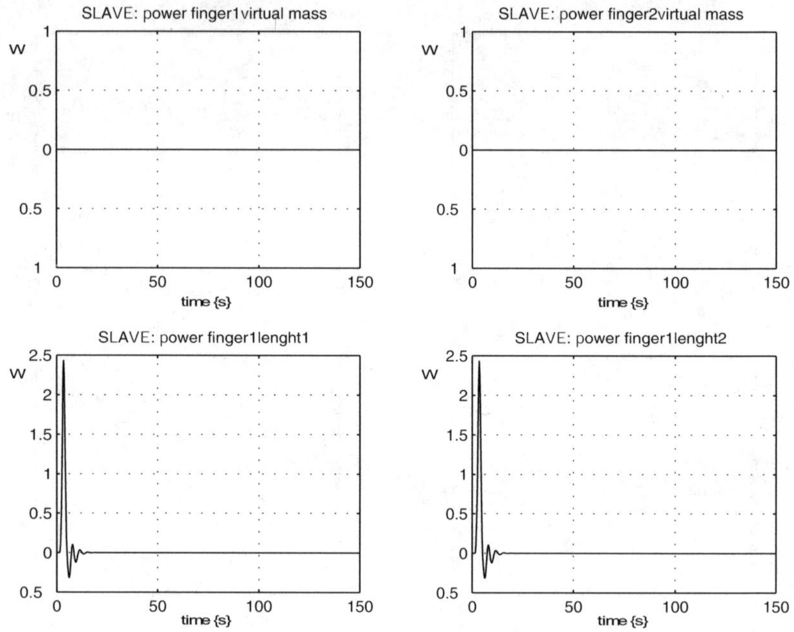

Fig. 4.11. The power at slave side

by the IPC proposed in Sec. 3.3.2. A port-Hamiltonian model of the controlled system is given by:

$$\begin{cases} \dot{x} = [J(x) - R(x)]\dfrac{\partial H}{\partial x} + \begin{pmatrix} g & 0 & 0 \\ 0 & g_{l_1} & 0 \\ 0 & 0 & g_{l_2} \end{pmatrix} \begin{pmatrix} f \\ f_{l_1} \\ f_{l_2} \end{pmatrix} \\ \begin{pmatrix} e \\ e_{l_1} \\ e_{l_2} \end{pmatrix} = \begin{pmatrix} g^T & 0 & 0 \\ 0 & g_{l_1}^T & 0 \\ 0 & 0 & g_{l_2}^T \end{pmatrix} \frac{\partial H}{\partial x} \end{cases} \tag{4.23}$$

The power port (e, f) is associated to the virtual point of the IPC and it allows to inject energy to move the whole gripper, while the power ports (e_{l_1}, f_{l_1}) and (e_{l_2}, f_{l_2}) are associated to the rest lengths of the fingers and allow to inject energy in order to change the relative position of the fingers. In order to be able to use this controlled gripper in a bilateral telemanipulation scheme it is necessary to add the feedthrough terms that allow to remove the wave reflection phenomenon. Thus, the new controlled gripper turns out to be:

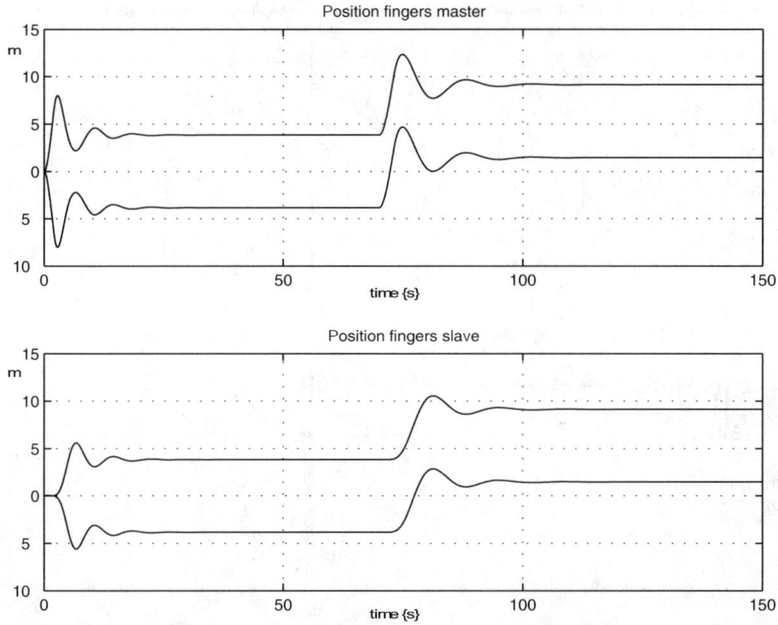

Fig. 4.12. Position of the fingers

$$\begin{cases} \dot{x} = [J(x) - R(x)]\dfrac{\partial H}{\partial x} + \begin{pmatrix} g & 0 & 0 \\ 0 & g_{l_1} & 0 \\ 0 & 0 & g_{l_2} \end{pmatrix} \begin{pmatrix} f \\ f_{l_1} \\ f_{l_2} \end{pmatrix} \\ \begin{pmatrix} e \\ e_{l_1} \\ e_{l_2} \end{pmatrix} = \begin{pmatrix} g^T \\ g^T_{l_1} \\ g^T_{l_2} \end{pmatrix} \frac{\partial H}{\partial x} + \begin{pmatrix} B(x) & 0 & 0 \\ 0 & B_{l_1}(x) & 0 \\ 0 & 0 & B_{l_2}(x) \end{pmatrix} \begin{pmatrix} f \\ f_{l_1} \\ f_{l_2} \end{pmatrix} \end{cases} \tag{4.24}$$

If the impedance of the scattering based communication channels used to transmit the power of the virtual point, and of the springs rest length ports are Z, Z_{l_1} and Z_{l_2} respectively, the power reflection phenomenon is removed if and only if $B(x) = Z$, $B_{l_1}(x) = Z_{l_1}$ and $B_{l_2}(x) = Z_{l_2}$.

In order to validate the telegrasping scheme proposed, some simulations are provided. Consider a simple telegrasping system in which each gripper has two fingers with one DOF. Energy relative to the rest lengths and virtual point power ports is transmitted through scattering based communication channels and line dampers have been added to the IPCs controlling each gripper. The mass of each finger is $m = 1\ Kg$; both the stiffness of the two variable springs and the one of the interaction spring is equal to $1\ N/m$. The mass of the virtual object is $m_c = 0.1\ Kg$ and the damper is $b = 1\ Nsec/m$. The transmission line has got a delay of $1\ sec$ and the line impedance is equal to 1. The first simulation deals with a finger opening task. Some force has been applied to

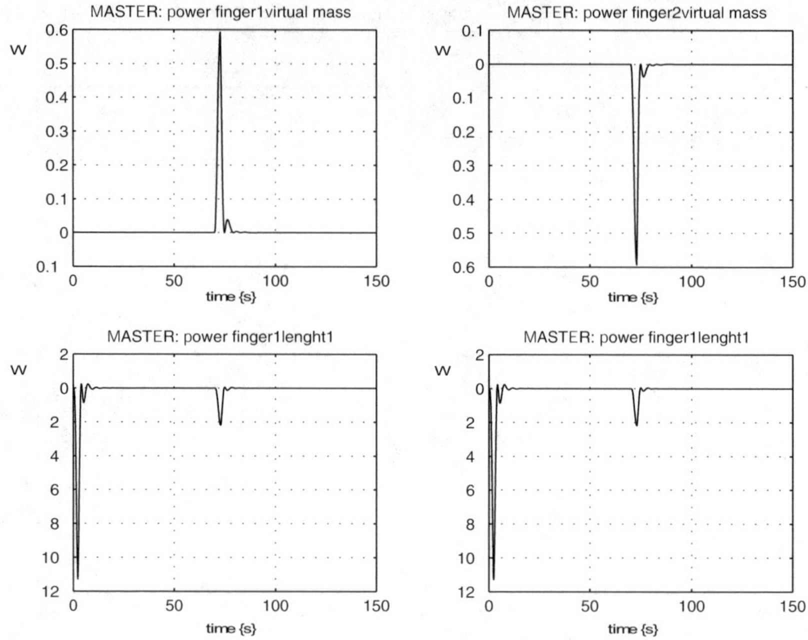

Fig. 4.13. The power at master side

the fingers of the master gripper in order to open the hand. All the energy flowing through the master power ports is transmitted to the slave side. The situation is illustrated in Fig. 4.9. Performances are very good and the slave follows exactly what the master does. Furthermore, the behavior of the overall system is stable, despite of the non negligible transmission delay. It is very interesting analyze the power graphs relative to the system. From Fig. 4.10 and Fig. 4.11 we can see that all the energy supplied by the supervisor is used to change the rest length of the variable spring opening in this way the two grippers. In the next simulation the master gripper has been first opened and then moved in the workspace. The situation is illustrated in Fig. 4.12. Analyzing the power flow in the system it can be seen that, when the master gripper is moved in the workspace, there is also a flow of energy through the virtual point power port. The situation is illustrated in figure Fig. 4.13 and figure Fig. 4.14. The last simulation is shown in Fig. 4.15. In this case the master gripper is opened and then closed but, while the master gripper is being closed, the slave touches an obstacle, placed at a distance $d_2 = 0.02\ m$ from finger 2, and at distance $d_1 = -0.05\ m$ from finger 1. It is possible to see that there is a good force reflection since the master, on which a force is applied to close the fingers, stops after the obstacle is touched at the remote side. It is also possible to see that, when a zero force is applied at the master,

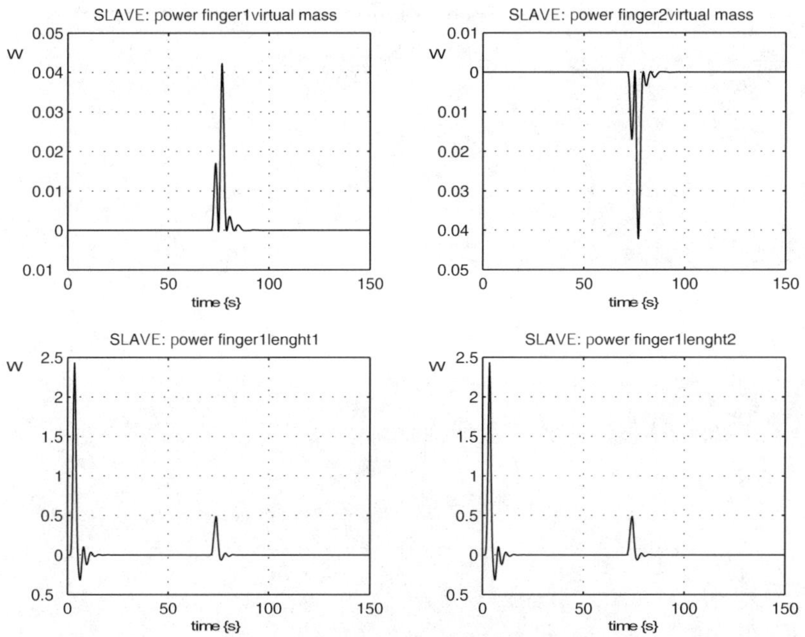

Fig. 4.14. The power at slave side

the two devices assume the same configuration because of the energy reflected from the slave side.

4.4 A Digital Scheme for Intrinsically Passive Telemanipulation

The intrinsically passive port-Hamiltonian based telemanipulation scheme proposed in Sec. 4.2 considers continuous time port-Hamiltonian controllers and communication channels characterized by a constant delay. In this section we show how to extend the scheme in order to take into account the digital nature of controllers. Furthermore, since a very appealing and cheap communication channel is the Internet, the energetic behavior of a packet switching transmission line is studied. There are two main problems to address: the strongly variable time delay and the possible unreliability due to the loss of packets associated with the channel.

The sampled data nature of the controllers can be considered by using the concepts described in Sec. 3.4. In fact, we can use the discretization algorithm reported in Sec. 3.4.1 in order to obtain a discrete port-Hamiltonian controller and to interconnect it by means of the energetic consistent Sample & Hold strategy described in Sec. 3.4.2 to the continuous plant. Since now the con-

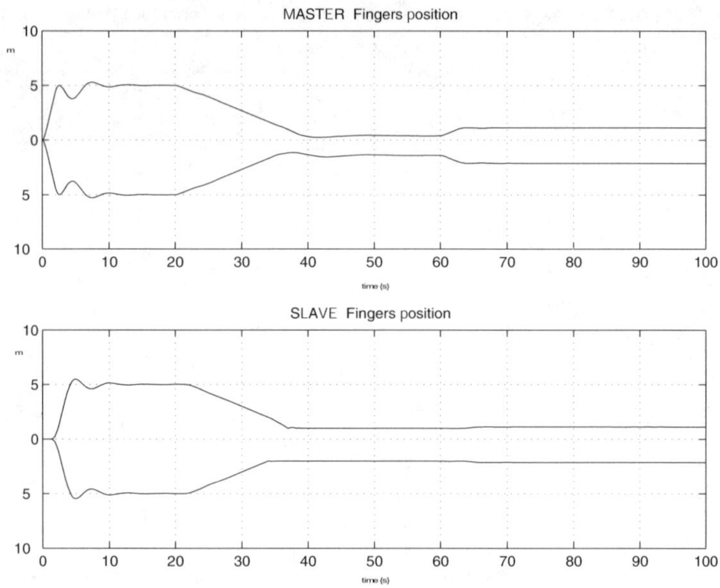

Fig. 4.15. Position of the fingers in a grasp.

trollers connected to the transmission line are discrete we have to consider a digital transmission line (e.g. Internet) through which master and slave sides will exchange energy. The proposed scheme is illustrated in Fig. 4.16 in a bond-graph notation. The symbols $\mathbb{C}$ and $\mathbb{R}$ represent the energy storing and energy dissipating part of each port-Hamiltonian system, both for the continuous plant and for the discrete controller. The barred bonds represent a discrete energy exchange.

Since we want to consider an Internet-like communication line, we have to consider a discrete communication channel and we will implement it by using scattering theory. It is possible to define a discrete time scattering. Each discrete time power port (either the master or the slave one) of the discrete communication channel illustrated in Fig. 4.17 is characterized by an effort $e(k)$ and by a flow $f(k)$. The energy flowing into the system from each port in one sample period is equal to:

$$Te^T(k)f(k)$$

We can always decompose the power flow into an incoming power wave and an outgoing power wave as shown in Theorem 1.35 and, by discrete integration, the energy flowing during one sample period from each power port is:

$$H(k) = Te^T(k)f(k) = \frac{T}{2}\|s^+(k)\|^2 - \frac{T}{2}\|s^-(k)\|^2$$

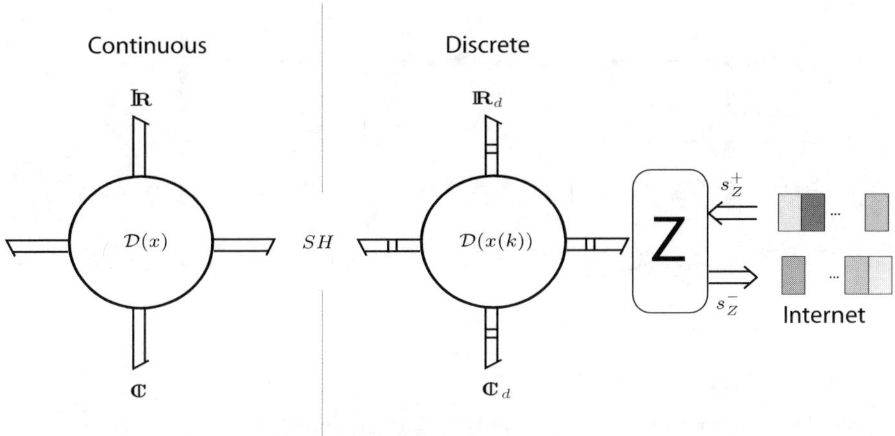

Fig. 4.16. The Passive Sampled Data Telemanipulation Scheme.

where T is the sample period. Thus we can interpret $\frac{T}{2}\|s^+(k)\|^2$ and $\frac{T}{2}\|s^-(k)\|^2$ as incoming and an outgoing energy packets respectively.

The discrete port-Hamiltonian controller is interconnected to the discrete scattering based communication channel as described in Sec. 4.2.4: at each sample time the system will acquire the incoming energy quantum $\frac{T}{2}s^+(k)$ and the discrete effort $e(k)$ and will calculate the discrete flow $f(k)$ and the discrete energy quantum $\frac{T}{2}s^-(k)$ to transmit through the communication channel. It is possible to compute $s^-(k)$ and $f(k)$ from $s^+(k)$ and $e(k)$ using Eq.(4.14a) and Eq.(4.14b).
We will use the following notation for the discrete derivative and the discrete integral:

$$dg(k) = \frac{g(k+1) - g(k)}{T} \qquad I_h^k g = \sum_{i=h}^{k-1} g(i)T \tag{4.25}$$

where g is a generic sequence. By trivial computations it can be shown that, in general:

$$dI_{k-h}^k g = g(k) - g(k-h) \tag{4.26}$$

In the following considerations we will assume that the protocol (e.g. the Internet protocol) used to implement the communication line doesn't change the order of the packets, such as the Internet TCP/IP protocol.
Let δ_{ms} and δ_{sm} be the delays associated to the communication between master and slave and slave and master respectively. The following result can be proven:

Proposition 4.4. *In case of fixed transmission delays and no loss of packets, the discrete communication channel is lossless in a discrete sense.*

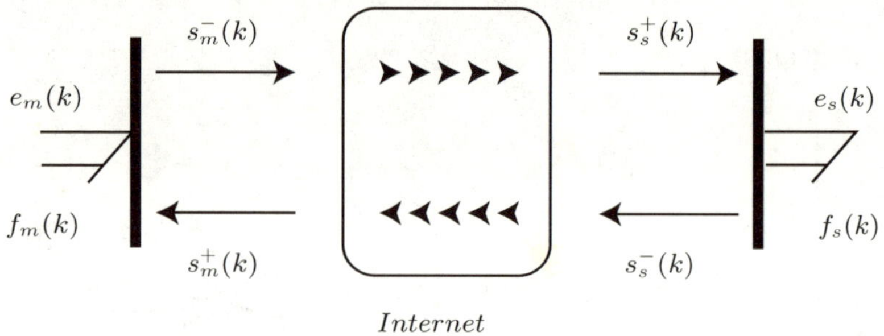

Fig. 4.17. The scattering based communication channel

Proof. The power flow into the communication channel is:

$$P(k) = \tfrac{1}{2}\|s_m^-(k)\|^2 + \tfrac{1}{2}\|s_s^-(k)\|^2 - \tfrac{1}{2}\|s_m^+(k)\|^2 - \tfrac{1}{2}\|s_s^+(k)\|^2 \tag{4.27}$$

but

$$s_s^+(k) = s_m^-(k - \delta_{ms})\ s_m^+(k) = s_s^-(k - \delta_{sm}) \tag{4.28}$$

Using Eq.(4.26), we can write

$$P(k) = \tfrac{1}{2}\|s_m^-(k)\|^2 - \tfrac{1}{2}\|s_m^-(k - \delta_{ms})\|^2 + \tfrac{1}{2}\|s_s^-(k)\|^2 - \tfrac{1}{2}\|s_s^-(k - \delta_{sm})\|^2 \tag{4.29}$$

We can, therefore, write:

$$P(k) = d[I_{k-\delta_{ms}}^{k}(\frac{1}{2}\|(s_m^-)\|^2) + I_{k-\delta_{sm}}^{k}(\frac{1}{2}\|(s_s^-)\|^2)] \tag{4.30}$$

Defining as energy function of the communication channel

$$H(k) = [I_{k-\delta_{ms}}^{k}(\frac{1}{2}\|s_m^-\|^2) + I_{k-\delta_{sm}}^{k}(\frac{1}{2}\|s_s^-\|^2)] \tag{4.31}$$

we have that:

$$P(k) = dH(k) \tag{4.32}$$

which means that all the power flowing through the communication channel is stored and that, therefore, the communication channel is lossless in a discrete sense.

In Internet-like communication channels some packets could be lost during transmission because of some traffic problems or some troubles in the servers each packet has to cross. Let us investigate the energetic behavior of the communication channel in the case one packet is lost in the transmission between master and slave. We propose to adopt the following *Replace with a Null packet* (RNP) strategy: when a packet is not delivered on time, it is replaced by a null packet. Thus, in case a packet is lost, we have that a 0 signal is received at the slave side instead of the correct value, namely $s_m^-(k-\delta_{ms})$ and that, therefore:

$$P(k) = \frac{1}{2}\|s_m^-(k)\|^2 + \frac{1}{2}\|s_s^-(k)\|^2 - \frac{1}{2}\|s_m^+(k)\|^2 \tag{4.33}$$

We can write:

$$\begin{aligned} P(k) &= \tfrac{1}{2}\|s_m^-(k)\|^2 + \tfrac{1}{2}\|s_s^-(k)\|^2 - \tfrac{1}{2}\|s_m^+(k)\|^2 = \tfrac{1}{2}\|s_m^-(k)\|^2 + \tfrac{1}{2}\|s_s^-(k)\|^2 - \\ &-\tfrac{1}{2}\|s_s^-(k-\delta_{sm})\|^2 - \tfrac{1}{2}\|s_m^-(k-\delta_{ms})\|^2 + \tfrac{1}{2}\|s_m^-(k-\delta_{ms})\|^2 = \\ &= dH(k) + \tfrac{1}{2}\|s_m^-(k-\delta_{ms})\|^2 \end{aligned} \tag{4.34}$$

We have an additional term which is positive definite and which represents a dissipated power. When a packet is lost in the transmission the channel dissipates an energy quantum corresponding to the value that it should have received from the master side. The behavior of the system, nonetheless, keeps on being passive.
Let, now, $\mathcal{P}_{ms}$ and $\mathcal{P}_{sm}$ be the set of time instants in which a packet in the communication between master and slave is lost and the set of time instants in which a packet in the communication between slave and master is lost respectively. The energetic behavior of the channel is:

$$P(k) = dH(k) + \alpha(\tfrac{1}{2}\|s_m^-(k-\delta_{ms})\|^2) + \beta(\tfrac{1}{2}\|s_s^-(k-\delta_{sm})\|^2) \tag{4.35}$$

where

$$\alpha = \begin{cases} 0 & k \notin \mathcal{P}_{ms} \\ 1 & k \in \mathcal{P}_{ms} \end{cases} \quad \beta = \begin{cases} 0 & k \notin \mathcal{P}_{sm} \\ 1 & k \in \mathcal{P}_{sm} \end{cases}$$

where α and β are coefficients that activate a power dissipation of the channel when a packet is lost.

Let us now consider the case in which the delay between master and slave is variable. Suppose that h is the minimum delay of the transmission between master and slave. Because of the variable delay, it can happen that one packet is so much delayed that the receiving queue is empty for some sample periods. In this case we can always write:

$$\begin{aligned} P(k) &= \tfrac{1}{2}\|s_m^-(k)\|^2 + \tfrac{1}{2}\|s_s^-(k)\|^2 - \tfrac{1}{2}\|s_m^+(k)\|^2 = \\ &= \tfrac{1}{2}\|s_m^-(k)\|^2 + \tfrac{1}{2}\|s_s^-(k)\|^2 - \tfrac{1}{2}\|s_s^-(k-\delta_{sm})\|^2 - \tfrac{1}{2}\|s_m^-(k-h)\|^2 + \\ &+\tfrac{1}{2}\|s_m^-(k-h)\|^2 = dH(k) + \tfrac{1}{2}\|s_m^-(k-h)\|^2 = dH(k) + P_d(k) \end{aligned} \tag{4.36}$$

where now we define:

$$H(k) = [I^k_{k-h}(\frac{1}{2}(s^-_m)^2) + I^k_{k-\delta_{sm}}(\frac{1}{2}(s^-_s)^2)] \tag{4.37}$$

as the energy function. We dissipate the energy quantum corresponding to the packet we were expecting. At a time $k+j$ the system receives the packet whose corresponding energy quantum was dissipated at time k. In this case it will be:

$$\begin{aligned} P(k+j) &= dH(k+j) - \tfrac{1}{2}\|s^-_m(k-h)\|^2 + \tfrac{1}{2}\|s^-_m(k+j-h)\|^2 = \\ &= dH(k+j) + P_p(k+j) + P_d(k+j) \end{aligned} \tag{4.38}$$

At sampling period $k+j$, the delayed incoming packet injects an extra energy $TP_p(k+j)$, on the other hand the missed packet $\frac{1}{2}\|s^-_m(k+j-h)\|^2$, since we are assuming that only one packet can be transmitted per each sample period, causes the dissipation of $TP_d(k+j)$ energy. We can see that the delay of a packet first implies a dissipation of an energy quantum and then an energy injection of the same amount of energy. It is, then, clear that

$$\sum_{n=0}^{\infty} P_p(n) + P_d(n) = 0$$

and, therefore, there is no global production of extra energy.
Both in case of fixed and variable delay the discrete scattering based communication channel is lossless; the main difference is that when we have a constant delay energy is neither produced nor dissipated but simply stored, while when we have a variable delay the energy quanta associated to the delayed packets are first dissipated and then injected back to the system. Since the variable delay introduces a *finite* extra-delay on the packets, passivity is preserved. One could expect that the energy injection leads to some non passive (and therefore potentially unstable) behavior but this is not the case since before being injected the extra energy has already been dissipated. In case some packets are lost the behavior of the communication channel is dissipative since we do not have any energy injection to recover the dissipated energy packets. It is now straightforward to state the following

Proposition 4.5. *Using the RNP strategy, the discrete scattering based communication channel is passive even in case of variable time delay and of loss of packets.*

Another possible way for dealing with unreliable networks, is the so called *Hold the Last Sample* (HLS) strategy, where, in case a packet is not received on time, it is replaced by the value of the packet that has been previously received. This strategy is not suitable for being used in port-Hamiltonian based telemanipulation because it can render the communication channel non passive in case of packet loss. In fact, suppose again that a packet is lost in the communication between master and slave sides. Using the HLS strategy, we have that at time k the non received packet $s_m^-(k-\delta_{ms})$ is replaced with the previously received one, namely $s_m^-(k-\delta_{ms}-1)$ and that, therefore:

$$P(k) = \frac{1}{2}\|s_m^-(k)\|^2 + \frac{1}{2}\|s_s^-(k)\|^2 - \frac{1}{2}\|s_s^-(k-\delta_{sm}\|^2 - \frac{1}{2}\|s_m^-(k-\delta_{ms}-1)\|^2 \tag{4.39}$$

we can always write

$$P(k) = \frac{1}{2}\|s_m^-(k)\|^2 + \frac{1}{2}\|s_s^-(k)\|^2 - \frac{1}{2}\|s_s^-(k-\delta_{sm}\|^2 - \frac{1}{2}\|s_m^-(k-\delta_{ms})\|^2 +$$
$$+\frac{1}{2}\|s_m^-(k-\delta_{ms})\|^2 - \frac{1}{2}\|s_m^-(k-\delta_{ms}-1)\|^2 =$$
$$= dH(k) + (\frac{1}{2}\|s_m^-(k-\delta_{ms})\|^2 - \frac{1}{2}\|s_m^-(k-\delta_{ms}-1)\|^2) \tag{4.40}$$

where $H(k)$ is defined as in Eq.(4.37).
The term term between brackets can be negative if the energy content of the lost packet is lower than that of the last packet received and, in this, case there is production of extra energy and, therefore, the communication channel has a non passive behavior. The passivity destructive behavior of the HLS strategy is magnified in case more packets are lost in the communication between master and slave and viceversa and in case of variable delay.
In order to illustrate the validity of the results obtained in this section, some simulations are proposed on a simple one degree of freedom telemanipulation system. Each robot (master and slave) is a 1 DOF system, a simple mass and is controlled by a sampled IPC, connected in a power consistent way to the continuous robot. The communication channel is implemented by means of the digital scattering strategy and the RNP strategy is used for dealing with packets loss and variable delay. In the first simulation we applied an impulsive force to the master and we plotted the positions of master and slave; the delay is constant and it is equal to 0.5 seconds. The results of this simulation are shown in Fig. 4.18. We can see that the smaller is the sample period, the closer is the behavior of the digital scheme to the continuous one, meaning that the discretization algorithm is well posed. Moreover, we can notice that the bigger is the sample time the worse are the performances. This is because the information we are transmitting gets worse; nevertheless the behavior of the overall system is passive independently of the value of the sample period.
In the next simulation we are considering a communication channel which represents the Internet. The delay in the communication between master and slave and between slave and master is variable (with a mean of 0.5 s.) and in the communication between master and slave every 2 seconds 10 packets are lost while in the communication between slave and master each 3 seconds 20 packets are lost. Once again we applied an impulsive force to the master

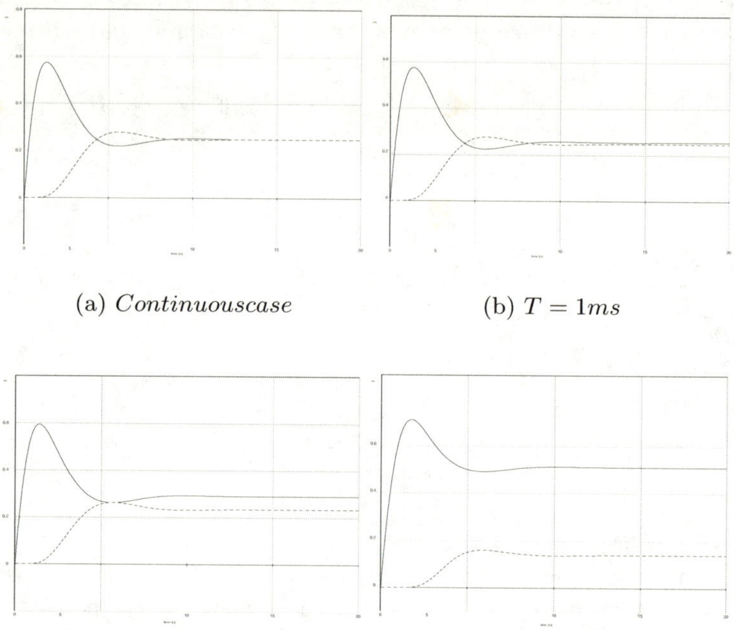

(a) *Continuouscase* (b) $T = 1ms$

(c) $T = 10ms$ (d) $T = 100ms$

Fig. 4.18. Free motion with different sample times. Position of the master (continuous) and of the slave (dashed)

and we plotted the positions of master and slave. The results are shown in figure Fig. 4.19. The dashed line represents the position of master and slave in case of constant delay (0.5 seconds) and of no loss of packets. The sample time is 10 ms. We can notice that the performances decrease but the stability is maintained because of the intrinsic passivity of the overall scheme. In the next simulation we implemented an interaction task. The master is pushed with a constant force and the slave interacts with a wall (implemented with a spring-damper system) posed at position $x = 0.1$ m. There is variable delay and loss of packets in the communication channel in both senses. The position of master and slave are shown in figure Fig. 4.20. We can see that when the slave stops when it touches the wall. The force of interaction is reflected back to the master side and compensates the force applied to the master. In fact we can see that the position of the master is constant even if the operator is keeping on applying a force.

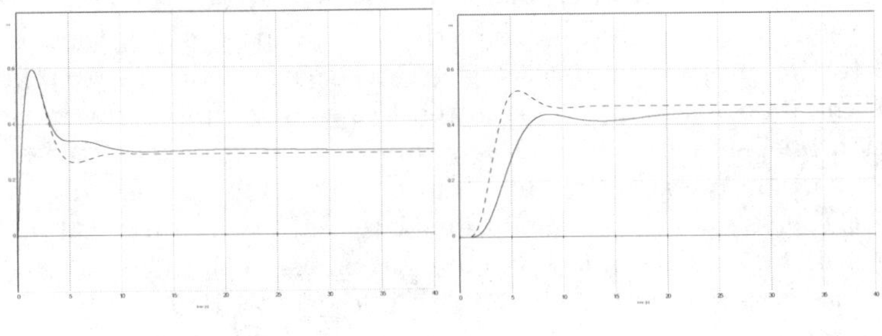

(a) Position of the Master (b) Position of the Slave

Fig. 4.19. Variable delay and loss of packets

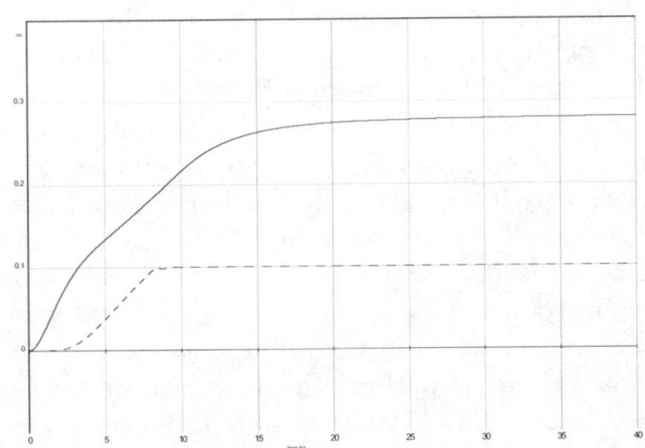

Fig. 4.20. Position of the master (continuous) ad of the slave (dashed)

4.5 Improving Performances in Intrinsically Passive Digital Telemanipulation

When using the RNP strategy, loss of packets in the discrete scattering based communication channels doesn't affect the passivity of the overall system but, nevertheless, it degrades the performances of the telemanipulation scheme since part of the energy necessary to complete a certain task is not transmitted but dissipated. Since in scattering based transmission lines the transmitted information represents a power wave, the packets are not totally uncorrelated and it does make sense to obtain the missed packets by interpolating the received ones. In this section we introduce a control scheme that is able to interpolate the received packets for reconstructing the missing ones, and, consequently, improving performances. Owing to keep the method simple, we focus on linear interpolation algorithm in our control strategy.

When we produce an interpolated packet we replace the energy content of the missing packet with the one of the interpolated one. If the energy associated to the interpolated packet is greater than the one associated to the lost one, the interpolation process introduces some extra energy in the communication line, similarly to what can happen when using the HLS strategy. The interpolation process, therefore, must be carefully addressed in order to preserve the passivity. By means of loss rate and other statistical indicators deriving from an analysis of the communication channel, we know which is, in average, the maximum number of consecutive packets that can be lost in the transmission and, consequently, the maximum number of consecutive packets that have to be interpolated; assume that this number is n. To the aim of performing a passive interpolation process, we need to endow the communication channel with transmission and receiving buffers.

In order to avoid energy production in the interpolation process, we need to know the maximum amount of energy that can be used to perform the possible interpolation of missed packets. The aim of the transmission buffer is to endow each packet with this extra information: the total energy H^a of the next n packets that will be subsequently transmitted. The transmission buffer, therefore, will introduce an extra delay on $n+1$ sample periods in order to endow each packet with the extra information required. The interpolation process will be performed in the receiving buffer and, since the maximum number of packets to be interpolated is n, its dimension will be, therefore, $n+2$ and it will introduce a delay of $n+2$ sample periods. Assume that we lose n packets, namely we receive nothing between $t = (k-n-1)T$ and $t = (k-1)T$. At $t = kT$ we can perform the interpolation between the received packets $s^+(k-n-2)$ and $s^+(k)$ (both present in the receiving buffer) to obtain the lost packets. In order to avoid any extra energy production, we use the information $H^a(k-n-2)$ embedded in the packet received in $t = (k-n-2)T$. We calculate the maximum energetic content ϵ that each packet can have by:

$$\epsilon = \frac{H^a(k-n-2)}{n} \tag{4.41}$$

Then we tune each interpolated packet in order to meet the energetic constraints.

Remark 4.6. Since there is not any information on the distribution of the energy among the lost packets, we fix the same energy bound for each packet to interpolate.

We can write the following algorithm:

1. Read $H^a(k-n-2)$, the energy available for the interpolation
2. Calculate the energetic content of each interpolated energy quantum: $\epsilon = \frac{H^a(k-n-2)}{n}$

3. By linear interpolation between $s^+(k)$ and $s^+(k-n-2)$ obtain $\bar{s}_I^+(i)$, with $i \in [k-n-1, k-1]$.
4. Obtain the packets to replace the missed ones, keeping into account energetic constraints:
   a) if $\frac{1}{2}\|\bar{s}_I^+(i)\|^2 T > \epsilon$ then choose $\alpha = \sqrt{\frac{2\epsilon}{\|\bar{s}_I^+(i)\|^2 T}}$
   else
   $\alpha = 1$
   end
   b) $s_I^+(i) = \alpha \bar{s}_I^+(i)$

   The total amount of energy H_I of the interpolated packets is:

$$H_I = \sum_{j=k-n-1}^{k-1} \frac{1}{2}\|s_I^+(j)\|^2 \leq H^a = \sum_{j=k-n-1}^{k-1} \frac{1}{2}\|s^+(j)\|^2$$

The energy introduced in the communication channel by the interpolation process is bounded by the energy content of the lost packets; no extra energy is introduced and, therefore, passivity is, intuitively, preserved.

Remark 4.7. If $m < n$ packets get lost it is possible to recover from the energy content of the $m-n$ received packets the energy available for the interpolation of the missed packets. If $p > n$ packets get lost, the interpolation algorithm fails and nothing is done to replace the lost packets; nevertheless, passivity is preserved and, in this case, the communication channel dissipates the energy associated to the missed packets.

The communication channel we are using for reliable telemanipulation is represented in Fig. 4.21. TX_m and TX_s represents the transmission buffers at master and slave side respectively and RX_m and RX_s the receiving buffers at

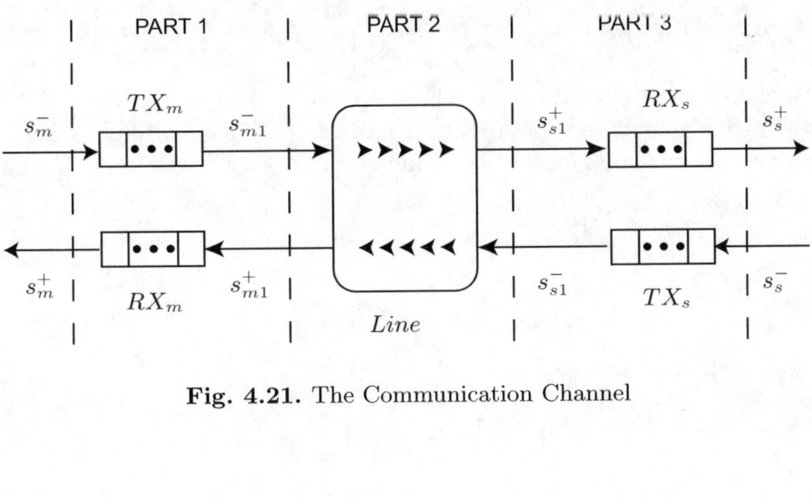

Fig. 4.21. The Communication Channel

master and slave side respectively. Let δ_{ms} and δ_{sm} the transmission delays between master and slave and slave and master respectively. Moreover, let δ_{Tm} and δ_{Ts} be the delays introduced by the transmission buffer at master and slave side respectively and δ_{Rm} and δ_{Rs} the delays introduced by the receiving buffer at master and slave side respectively. We now formally prove that the reliable channel proposed is passive.

Suppose that 1 packet is lost in the transmission between master and slave. We will use the same notation reported in Eq.(4.25) for the discrete derivative and the discrete integral. The communication channel can be divided in three parts as shown in Fig. 4.21. Consider the first part of the communication channel; we have that the power flow is:

$$P_1(k) = \frac{1}{2}\|s_m^-(k)\|^2 - \frac{1}{2}\|s_{m1}^-(k)\|^2 + \frac{1}{2}\|s_{m1}^+(k)\|^2 - \frac{1}{2}\|s_m^+(k)\|^2 \tag{4.42}$$

Since there is no loss of packets between slave an master, we have:

$$\begin{cases} s_{m1}^-(k) = s_m^-(k - \delta_{Tm}) \\ s_m^+(k) = s_{m1}^+(k - \delta_{Rm}) \end{cases}$$

and, therefore, if we pose:

$$H_1(k) := [I_{k-\delta_{Rm}}^k(\frac{1}{2}\|(s_{m1}^+)\|^2) + I_{k-\delta_{Tm}}^k(\frac{1}{2}\|(s_m^-)\|^2)]$$

we have that:

$$P_1(k) = dH_1(k)$$

and therefore the first part of the communication channel is lossless.

Let us now consider the second part of the communication channel; we have that:

$$P_2(k) = \frac{1}{2}\|s_{m1}^-(k)\|^2 - \frac{1}{2}\|s_{s1}^+(k)\|^2 + \frac{1}{2}\|s_{s1}^-(k)\|^2 - \frac{1}{2}\|s_{m1}^+(k)\|^2 \tag{4.43}$$

As proven in Proposition 4.5, when a packet is lost and strategy RNP is adopted, the channel becomes dissipative instead of lossless. Suppose that we do not receive anything at the slave side at time $t = kT$. Defining

$$H_2(k) = [I_{k-\delta_{ms}}^k(\frac{1}{2}\|(s_{m1}^-)\|^2) + I_{k-\delta_{sm}}^k(\frac{1}{2}\|(s_{s1}^-)\|^2)]$$

we have that

$$P_2(k) = dH_2(k) + P_{d2}(k)$$

We have a dissipative behavior and

$$P_{d2}(k) = \frac{1}{2}\|s_{m1}^-(k - \delta_{ms})\|^2 > 0$$

and we dissipate an energy amount equal to:

$$H_{d2}(k) = \frac{1}{2}\|s_{m1}^{-}(k-\delta_{ms})\|^2 T$$

Let us now consider the third part of the communication channel. Since one packet is lost in the transmission between master and slave, suppose nothing is received at $t = kT$ at the slave side. We have that the power $P_3(k)$ flowing in the third part of the communication channel is:

$$P_3(k) = -\frac{1}{2}\|s_s^{+}(k)\|^2 + \frac{1}{2}\|s_s^{-}(k)\|^2 - \frac{1}{2}\|s_{s1}^{-}(k)\|^2$$

We can write:

$$P_3(k) = \frac{1}{2}\|s_{s1}^{+}(k)\|^2 - \frac{1}{2}\|s_s^{+}(k)\|^2 + \frac{1}{2}\|s_s^{-}(k)\|^2 -$$

$$-\frac{1}{2}\|s_{s1}^{-}(k)\|^2 - \frac{1}{2}\|s_{s1}^{+}(k)\|^2 = dH_3(k) - \frac{1}{2}\|s_{s1}^{+}(k)\|^2$$

where

$$H_3(k) = d[I_{k-\delta_{Rs}}^{k}(\frac{1}{2}\|(s_{s1}^{+})\|^2) + I_{k-\delta_{Ts}}^{k}\frac{1}{2}\|(s_s^{-})\|^2)]$$

We will have, therefore, an energy production at $t = kT$:

$$H_{p3} = P_{p3}(k)T = \frac{1}{2}\|s_{s1}^{+}(k)\|^2 T$$

At time $t = (k + \delta_{Rs})T = hT$ the lost packet is replaced with the interpolated one and, therefore, instead of delivering $s_s^{+} = 0$ for the missed packet, it will be delivered $s_s^{+} = s_{sI}^{+}$. When s_s^{+} is replaced with an interpolated packet, we have that:

$$P_3(h) = \frac{1}{2}\|s_{s1}^{+}(h)\|^2 - \frac{1}{2}\|s_{sI}^{+}(h)\|^2 + \frac{1}{2}\|s_s^{-}(h)\|^2 - \frac{1}{2}\|s_{s1}^{2}(h)\|$$

We can always write:

$$P_3(h) =$$

$$= \frac{1}{2}\|s_{s1}^{+}(h)\|^2 - \frac{1}{2}\|s_s^{+}(h)\|^2 + \frac{1}{2}\|s_s^{-}(h)\|^2 - \frac{1}{2}\|s_{s1}^{2}(h)\| + \frac{1}{2}\|s_s^{+}(h)\|^2 - \frac{1}{2}\|s_{sI}^{+}(h)\|^2$$

and, therefore,

$$P_3(h) = dH_3(h) + \frac{1}{2}\|s_s^{+}(h)\|^2 - \frac{1}{2}\|s_{sI}^{+}(h)\|^2 = dH_3(h) + P_{d3}(h)$$

By construction $P_{d3} > 0$ and, therefore, the interpolation algorithm introduces dissipation into the system. Now, we will put all the parts of the communication channel together in order to prove the overall passivity. We have that the power flowing through the communication channel is:

$$P(k) = \frac{1}{2}\|s_m^-(k)\|^2 - \frac{1}{2}\|s_s^+(k)\|^2 + \frac{1}{2}\|s_s^-(k)^2\| - \frac{1}{2}\|s_m^+(k)\|^2 = \\ = P_1(k) + P_2(k) + P_3(k) \tag{4.44}$$

When a packet is lost in the communication between master and slave, we have that:

$$P(k) = d(H_1(k) + H_2(k) + H_3(k)) + P_{d2}(k) - P_{p3}(k) = \\ = dH(k) + P_{d2}(k) - P_{p3}(k)$$

Since $P_{d2}(k) = P_{p3}(k)$ we have that

$$P(k) = dH(k)$$

and therefore the channel has a lossless behavior. At time $t = hT = (k+\delta_{Rs})T$ we have that:

$$P(h) = dH(h) + P_{d3}(h)$$

We can therefore conclude that the loss of a packet in the transmission line introduces a dissipative behavior in the communication channel. This result can be easily generalized both in case more than one packet is lost and in case the loss is in the communication between slave and master. We can therefore state the following:

Proposition 4.8. *If the RNP strategy is adopted and and the proposed interpolation algorithm is used, then the communication channel is passive.*

Remark 4.9. Both in case of interpolated and non interpolated communication channels, we obtain a dissipative communication channel. The improvement of performances obtained by means of the interpolation relies in the fact that the dissipated energy is much lower. This leads to a communication channel closer to the ideal case (i.e. losslessness) and therefore to better performances.

Remark 4.10. Both transmission and reception buffers introduce some extra delays necessary to endow packets with energetic information and to implement the interpolation. If the maximum number of possibly consecutive missed packet is quite small, the delay introduced is not so big and performances improvement is worth of it. In case bigger delays have to be introduced, it could be better, depending on the application, to maintain the dissipative behavior due to loss of packets rather than introducing a big extra delay.

In order to show the effectiveness of the interpolation algorithm, we simulated a simple 1-DOF telemanipulation system. The master and the slave are simple masses which are controlled by a discrete port-Hamiltonian controller and which communicate by means of a packet switching communication channel. The transmission delay is 0.5 seconds and the sample time is $T = 5msec$.

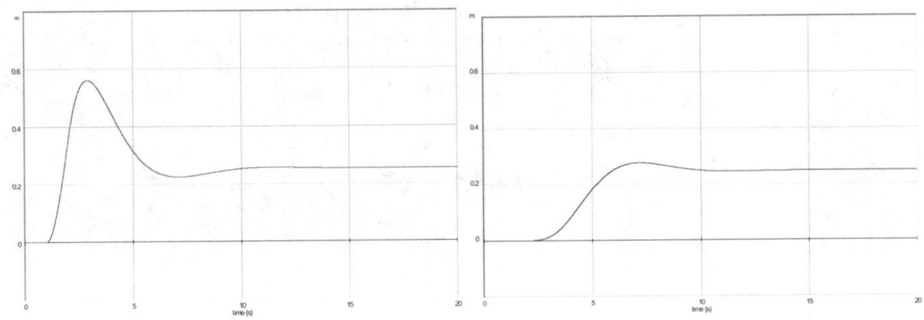

(a) Master Position (b) Slave Position

Fig. 4.22. Positions of Master and Slave

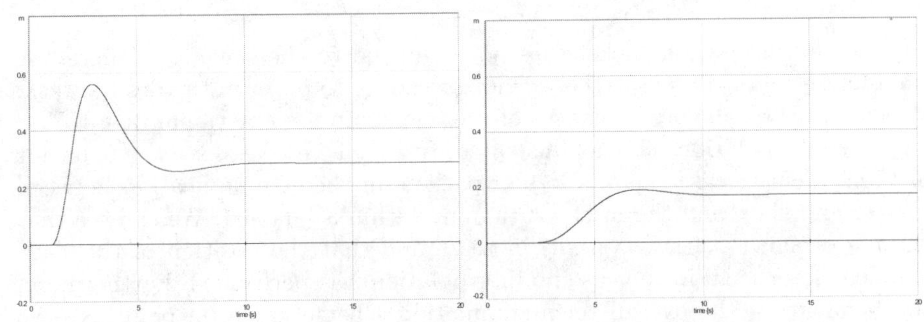

(a) Master Position (b) Slave Position

Fig. 4.23. Positions of Master and Slave in case of loss of packets

In Fig. 4.22 is represented the behavior of the system in case there is no loss of packets during the transmission. We can see that the slave follows the master after a certain delay.

Next we simulated a loss of packets in the communication between master and slave. The loss rate is 30%, namely one packet is lost per each three packets transmitted.

In Fig. 4.23 the behavior of the system is shown. We notice that the behavior of the system is stable since the loss of packets implies a dissipation of energy within the communication channel and, therefore, the passivity of the overall system is not compromised, as proven in Sec. 4.4. Nevertheless the performances are affected by unreliability of the transmission line. In fact, we can see that the slave position is quite different from the master one. This

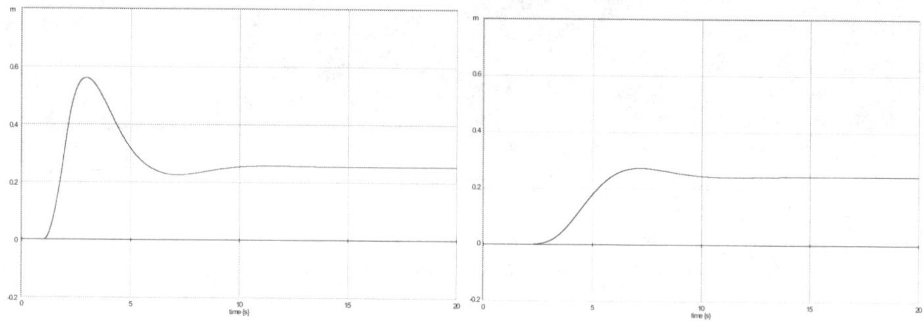

(a) Master Position (b) Slave Position

Fig. 4.24. Positions of Master and Slave in case of loss of packets and interpolation algorithm enabled

happens because a lot packets are not delivered to the slave and, therefore, the remote robot does not receive enough energy to perform the desired task. We can improve the performances of the telemanipulation system by introducing the interpolation scheme illustrated. In Fig. 4.24 we can see the behavior of the system in case there is loss of packets in the communication between master and slave and the interpolation algorithm is enabled. We can see that the performances of the systems increase and that the position of the slave is much closer than in the case no interpolation was performed. Furthermore, the behavior of the overall telemanipulation scheme keeps on being passive since the interpolation process in implemented in a passivity preserving way, as proven in Proposition 4.8.

4.6 Conclusions

A bilateral telemanipulation system is a robotic system that allows a human operator to interact with a remote environment. Passivity theory and the port-Hamiltonian formalism provide a framework for modeling and controlling complex, possibly nonlinear, telemanipulators. In fact, it is possible to model master and slave robots as port-Hamiltonian systems and to design intrinsically passive port-Hamiltonian regulators for controlling the behavior of the robots. The destabilizing effect of the delay in the communication between master and slave sides is solved by using the geometric scattering theory that allows to implement a communication channel which is lossless independently of any constant communication delay and that allows to transmit geometric quantities such as flows and efforts in a coordinate free way rather than simple scalars. The recent work on port-Hamiltonian based telemanipulation focuses on position tracking [269], force and velocity scaling [264, 265, 270] and on the use of predictive control techniques for improving performances [43].

Using the techniques reported in Chap. 3 it is possible to passively discretize the port-Hamiltonian controllers and to consider digital transmission lines as packet-switching networks (e.g. Internet). A discrete scattering based communication channel has been defined and a strategy for passively dealing with packets loss and variable delay has been developed. This allows to safely use Internet as a medium to interconnect master and slave sides. In [171], the HLS strategy is used for dealing with variable communication delay and a compensation algorithm for re-establishing passivity of the communication channel is developed. In [20] the methods proposed in [171] and in Sec. 4.4 are compared and a passivity preserving algorithm, alternative to that presented in Sec. 4.5, is proposed. In [48] the discrete scattering framework has been used for developing stable haptic display for slowly updated virtual environments.

5

Transparency in Port-Hamiltonian Based Telemanipulation

5.1 Introduction

Stability is a key issue in the implementation of a bilateral telemanipulation system since both the non-negligible time delay in the communication between master and slave and the interaction with unknown environments can destabilize the whole system. In Chap. 4 it has been shown that passivity theory and port-Hamiltonian systems can be fruitfully used to build an intrinsically passive telemanipulation scheme which, therefore, has a stable behavior both in case of free motion and in case of contact with any passive, possibly unknown, environment. Scattering theory has been used to build a communication channel which is lossless independently of any constant transmission delay and the problem of wave reflection arising when coupling master and slave side through scattering based communication channels has been solved. The scheme has been extended in order to take into account the sampled data nature of controllers in a passive way. Moreover discrete scattering has been defined and packet-switching transmission lines have been considered. A communication strategy that allows to preserve passivity even in case of loss of packets and of variable transmission delay has been proposed. Finally, a passivity preserving algorithm that allows to rebuild lost packets by interpolation has been proposed.

Intrinsic passivity, and, therefore, a stable behavior, is only a *first* step towards the implementation of a telemanipulation system. In fact performances have to be taken into account to make a telerobotic system effective and useful for real applications. After stability, *transparency* is the major goal in teleoperation systems design. A telemanipulator behaves as better as much the operator at the master side feels to interact directly with the remote environment. The whole telemanipulation system (robots, controllers and communication channel) should ideally be *transparent* and the operator should not feel its presence at all.

Several researches addressed the problem of transparency. In [339] transparency is defined as a correspondence between master and slave positions and forces while in [155] as a match between the impedance perceived by the operator and the environment impedance. Furthermore, several telemanipulation schemes that optimize transparency have been proposed. In [47] linear telemanipulators characterized by a negligible communication delay are considered and a generalized control configuration, merging a position feedback and a force feedback scheme, is developed to provide ideal transparency in terms of impedances match. In [94] a four channel teleoperator system is considered and the nonlinear dynamics of the master/human system is approximated by a linear dynamic model with varying parameters; a control strategy that optimizes the transparency of the overall approximated system is proposed. In [116] telemanipulators characterized by a non negligible time delay are considered and the effect of local force feedback on the transparency of the overall system is analyzed. In [93] frequency-domain loop-shaping techniques are used to provide transparency and stability in a bilateral telemanipulation system. Transparency analysis is carried on using linear control tools (e.g. Llewellyn's criterion for absolute stability) and linear or linearized telemanipulators are considered.

The aim of this chapter is to enhance transparency of the port-Hamiltonian based telemanipulation scheme proposed in Chap. 4. In order to be able to reproduce the remote environment behavior at the master side, it is crucial to choose a suitable way for representing the dynamics of the contact and, for the reasons explained in the next section, we will choose the Hunt-Crossley model.

The port-Hamiltonian framework allows to consider both linear and nonlinear bilateral telemanipulation systems and, therefore, in order to maintain this advantage, we need a definition and an evaluation method for transparency that is general enough and that does not rely on linear analysis tools. Thus, we introduce a general framework for the analysis of transparency based on behavioral approach [330, 295] and on the concept of *jet space* (see Sec. 1.2.3). The tools used for the definition of this framework are very general and can be used for the analysis of both linear and nonlinear telemanipulators. We then present a transparency analysis of the scattering based communication channel used in port-Hamiltonian based telemanipulation; we highlight the influence of packets loss and of the transmission delay on transparency and we propose a way to minimize it.

The impedance controller at master side strongly contributes to the feeling perceived by the human operator and, therefore, its virtual physical structure should mimic that of the remote environment the slave is interacting with and it should be possible to change the parameters of the controller to adapt it to the various kind of environment the slave can interact with. Thus, finally, we show how to "shape" the port-Hamiltonian controllers in order to optimize transparency while preserving passivity of the overall scheme.

5.2 A Model for the Contact Impedance

A transparent telemanipulation system has to reproduce at the master side exactly the same interactive behavior taking place between the slave and the remote environment which depends on the contact dynamics. In order to maximize the transparency of port-Hamiltonian telemanipulation systems, it is first of all necessary to have a target dynamics to reproduce at the master side or, in other words, to model the contact impedance.

Several ways for describing viscoelastic materials have been proposed in the literature, see [101] for a survey. The simplest way to represent a visco-elastic contact is the Kelvin-Voigt model, where the environment is represented as a parallel interconnection of a spring and a damper. This kind of model, because of its linearity, is not suitable for describing the behavior of soft materials, such as human tissues, where viscous effects are nonlinear [70, 294, 126]. In [126] a nonlinear model of the contact where the damping coefficient depends on the relative penetration of the contacting bodies is proposed. This description, known as the Hunt-Crossley model, gives a more accurate representation of the contact dynamics than the Kelvin-Voigt model and it is consistent with the notion of coefficient of restitution, used to characterize the energy loss during the contact [69, 70]. Thus, in order to preserve a proper degree of generality and to encompass a large class of environments, we model the contact impedance between the slave and the environment using the Hunt-Crossley model.

We will illustrate the Hunt-Crossley model for the one dimensional case to keep the notation simple; the generalization to the multidimensional case is straightforward. Consider the interaction of the slave robot with a remote environment posed in a certain position x_e. The environment can be modeled with an impedance causality, namely, it receives a velocity and provides a force. When using the Hunt-Crossley model, the impedance exhibited by the environment can be expressed by:

$$F(t) = \begin{cases} 0 & x < x_e \\ -kx^\alpha(t) - bx^\alpha \dot{x}(t) & x \geq x_e \end{cases} \tag{5.1}$$

where $F(t)$ is the force provided by the environment, x is the penetration into the object and $\dot{x}$ is the velocity of penetration; k and b are the elastic and the damping coefficients that characterize the contact respectively. The nonlinear dependence of the damping on the position is needed to faithfully model the force behavior when the robot is abandoning the contact. Finally, the coefficient α, which is usually close to one, models the dependence of the contact from the geometric properties of the interacting systems.

Thus, the contact behavior is completely known once the coefficients α, k and b are determined. These parameters can either be known a priori or they can be identified starting from force and position measurements. In particular, if the coefficient α is approximately known, since the contact model is linear with respect k and b, it is possible to estimate the elastic and the damping coefficients by means of the least squares method[138]. For a recent efficient contact impedance estimation algorithm see [69, 70].

5.3 A Behavioral Framework for Evaluating Transparency

In Chap. 1 we have seen that the behavioral approach introduced in [330, 243] is a very powerful tool for modeling and it has been shown that port-Hamiltonian framework fits very well into this modeling philosophy. Furthermore the behavioral approach has been very useful also from a control point of view, as illustrated in Chap. 2 where the control as interconnection paradigm allowed to have many more insights in the energy shaping problem.

In Chap. 3 it has been shown that interaction between physical systems takes place through localized power ports through which the systems exchange energy. Thus, in order to understand the way in which the human operator perceives the remote environment, it is necessary to describe the behavior of the system at these ports. The aim of this section is to exploit the behavioral paradigm to describe the behavior at the interaction power port and, hence, to develop a general framework for the study of transparency in bilateral telemanipulation systems, both linear and nonlinear.

5.3.1 Analysis of the Port Behavior

For a spatial manipulator with n links interacting with the environment, the configuration manifold $\mathcal{X}$ can be represented with the Cartesian product of the six-dimensional Lie group $SE(3)$; each Lie group represents the space of homogeneous matrices expressing the spatial configuration of each interacting link. Thus:

$$\mathcal{X} = \underbrace{SE(3) \times \cdots \times SE(3)}_{n\ times} \tag{5.2}$$

In case there is only one end effector interacting with the environment, $\mathcal{X} = SE(3)$.

The set of *interaction flows* for a robotic system interacting with n links is the following vector space:

$$F = \underbrace{se(3) \times \cdots \times se(3)}_{n\ times} \tag{5.3}$$

where $se(3)$ is the Lie algebra associated to $SE(3)$ and represents the set of twists at the interaction power ports of the robots. The set of *interaction efforts* is given by:

$$E = \underbrace{se^*(3) \times \cdots \times se^*(3)}_{n\ times} \tag{5.4}$$

where $se^*(3)$ is the dual space of the Lie algebra associated to $SE(3)$ and represents the set of wrenches that can be applied to the interacting part of the robot (see Sec. A.3). We can now define the *Port outcome space* as the product of the interaction flows and efforts spaces, namely:

$$\tilde{W} = F \times E \tag{5.5}$$

Both the space of interaction flows (twists) and the space of interaction efforts (wrenches) are independent of the configuration for the reasons illustrated in Sec. 1.3 and, therefore, the port outcomes space is a vector space independent of the configuration. In order to take into account configurations, an extended port outcomes space can be defined:

Definition 5.1 (Extended port outcome space). *The* extended port outcomes space *is the space:*

$$W = \tilde{W} \times \mathcal{X} \tag{5.6}$$

The extended port outcomes space depends on the configuration and it is NOT a vector space. Since F and E are given by the Cartesian product of dual vector spaces, they are obviously dual vector spaces and therefore it is possible to intrinsically define the instantaneous power of a port outcome;. Given $\tilde{w} = (f, e) \in \tilde{W}$, the instantaneous power of $\tilde{w}$, $\Pi(\tilde{w})$, is the dual product between e and f.

$$\Pi(\tilde{w}) = \langle e, f \rangle = \sum_{i=1}^{n} \langle e_i, f_i \rangle \; f_i \in se(3), e_i \in se^*(3) \tag{5.7}$$

The definition of power is independent of the configuration and represents the power the robot exchanges with the environment it is interacting with.

Let $\mathbb{R}$ be the time set and let $\mathcal{M}^{\mathbb{R}}$ indicate the set of maps form $\mathbb{R}$ to the space $\mathcal{M}$. It is possible to define the integral of a flow given an initial instant $t_0 \in \mathbb{R}$ and the corresponding initial configuration $x(t_0) \in \mathcal{X}$.

Definition 5.2 (Temporal integral of a flow). *The* temporal integral of a flow $F(\cdot) \in F^{\mathbb{R}}$ *over the time set* $\mathbb{R}$ *and with initial configuration* x_0*, denoted by* $\int_{\mathbb{R},x_0} F(\tau)d\tau$*, is the end-effector configuration* $x(\cdot) \in \mathcal{X}^{\mathbb{R}}$ *such that*

$$R_{x^{-1}(t)}(x(t), \dot{x}(t)) = (\mathbf{e}, F(t))$$

and $x(t_0) = x_0$*, where* $R^{-1}(\cdot)$ *denotes the Lie group right translation and* $\mathbf{e}$ *denotes the identity element of the Lie group* $SE(3) \times \cdots \times SE(3)$.

Using the right translation, the configuration results expressed in an inertial frame, while, if we used the left translation, the configuration would have been expressed in a frame rigidly connected to the power port. We have now all the elements necessary for giving the following:

Definition 5.3. *The* universe of port-outcomes *is the set*

$$\mathcal{U} = \{(x, f, e) \in W^T | R_{x^{-1}(t)}(x(t), \dot{x}(t)) = (\mathbf{e}, f(t)) \forall t \in T\} \tag{5.8}$$

The universe of port-outcomes represent all the outcomes that are compatible with the structure of the system under consideration. The *port behavior* can be then defined as $\mathcal{B} \subset \mathcal{U}$, namely as a subset of compatible port-outcomes. The behavior of a dynamical system can be represented through dynamical behavioral equations using the concept of jet space, see Sec. 1.2.3. In fact, it is possible to define the following

Definition 5.4 (extended n^{th}-order port jet space). *The extended n^{th}-order port jet space is defined to be the space $\mathbb{R} \times W^{(n)}$, where $W^{(n)} = \mathcal{X} \times \tilde{W}^{(n)}$ and $\tilde{W}^{(n)}$ is the nth prolongation of $\tilde{W}$, defined as in Eq.(1.4).*

Let $\mathcal{B}$ be a behavior defined as a subset of the universe of port outcomes. It is possible to express a particular behavior by means of a set of differential equations by using the extended n^{th} order jet space.

Definition 5.5 (Behavior representable by a differential equation). *The behavior $\mathcal{B} \subset \mathcal{U}$ can be represented by a differential equation if there exists a continuous function $\Delta_v^{\mathcal{B}} : \mathbb{R} \times W^{(n)} \to \mathbb{R}^v$, called an associated differential equation to $\mathcal{B}$, such that the subset of $\mathbb{R} \times W^{(n)}$ defined by:*

$$\mathcal{S}_{\Delta_v^{\mathcal{B}}} = \{(t, w^{(n)}) \quad | \quad \Delta_v^{\mathcal{B}}(t, w^{(n)}) = 0^v\} \tag{5.9}$$

is equal to $\mathcal{B}$.

The previous definition allows to express a behavior, a subset of the universe of port outcomes, as the kernel of a properly defined map on the nth-order extended jet space of the port outcomes.
Both for verification of performances and for control purposes it is useful to define the *distance* between two port behaviors. Suppose that a certain behavior is described by $\Delta_v^{\mathcal{B}}$ and that $w_m(t)$ is the measured port outcome at time t.

Definition 5.6 (Behavioral deviation). *The behavioral deviation at time t, $\varepsilon(t)$, of $w_m(\cdot)$ from a the behavior expressed by $\Delta_v^{\mathcal{B}}$ is defined to be*

$$\varepsilon(t) = \|\Delta_v^{\mathcal{B}}(t, pr^{(n)} w_m(t))\| \tag{5.10}$$

where $\|\cdot\|$ is the Euclidean norm of $\mathbb{R}^v$

Very often a behavior depends on a certain set of parameters which can be time varying. In the sequel, this set of parameters will be indicated as:

$$p(t) = (p_1(t), \cdots, p_n(t)) \quad t \in \mathbb{R} \tag{5.11}$$

and the dependence of a certain behavior on a set of parameters will be made explicit, when needed, by the following notation: $\Delta_v^{\mathcal{B},p(t)}(\cdot)$.

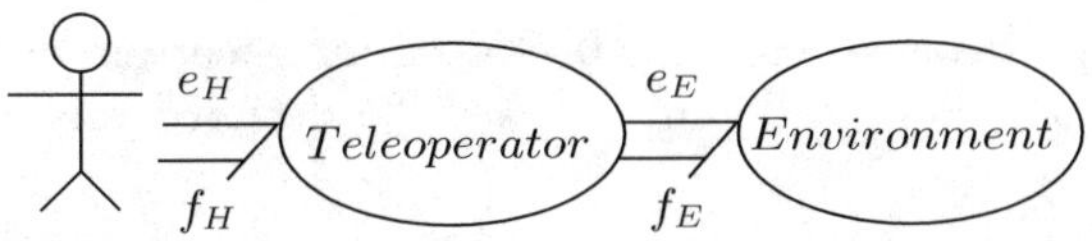

Fig. 5.1. A generic teleoperator.

5.3.2 Transparency in Telemanipulation

A port-Hamiltonian based telemanipulation system can be represented by the scheme in Fig. 5.1. Two ports can be distinguished: the power port (e_E, f_E) by means of which the teleoperator interacts with the environment and the power port (e_H, f_H) by means of which the human interacts with the telemanipulation system. Both ports are used to describe the interaction between some mechanical linkages (master and slave robots) and some external systems (the human operator and the environment). The human operator interacts with the links of the master robot which correspond to the links of the slave robots that interact with the remote environment. Thus, the two ports have the same extended port outcomes space (W, defined as in Eq.(5.6)) and the same universe of port outcomes ($\mathcal{U}$, defined in Eq.(5.8)).

A telemanipulation system is perfectly transparent if the behavior at the port (e_H, f_H) is the same as the behavior at the port (e_E, f_E); in fact, if this happened, it would mean that the feeling deriving by the interaction between the slave robot and the environment (determined by the *dynamic relation* between e_E and f_E) is exactly reproduced at the port (e_H, f_H) and, therefore, the operator would have the feeling of interacting with the environment as if he/she were directly doing that. Let $\mathcal{B}_e \in \mathcal{U}$ be the port behavior of the environment and $\Delta_v^{\mathcal{B}_e, e(t)} : \mathbb{R} \times W^{(n)} \to \mathbb{R}^v$ be the set of differential equations representing the port-behavior $\mathcal{B}_e$ depending on the set of parameters $e(t)$. Let, furthermore, $w_H(t) \in W$ be the port outcome at the human port at time t.

Definition 5.7 (Transparency). *A telemanipulation system is* transparent *if:*

$$\Delta_v^{\mathcal{B}_e, e(t)}(t, w_H(t)) = 0_v \tag{5.12}$$

Thus perfect transparency is achieved if the port-outcome perceived by the human exactly reproduces the behavior at the environment port. A measure of non transparency is given by the deviation of the human port from the behavior $\mathcal{B}_e$. The following *transparency deviation index* can be defined.

Definition 5.8 (Transparency deviation index). *The* transparency deviation index ε *is defined by the following relation:*

$$\varepsilon(t) = \|\Delta_v^{\mathcal{E}, e(t)}(t, pr^{(n)} w_H(t))\| \tag{5.13}$$

In case of perfect transparency $\varepsilon = 0$. To require a certain degree of transparency, it is possible to fix a bound $a \in \mathbb{R}^+$ and design a telemanipulator such that:

$$\varepsilon(t) \leq a \quad \forall t \in \mathbb{R} \tag{5.14}$$

Transparency and transparency deviation index as defined in Def. 5.7 and Def. 5.8 are very general concepts valid for both linear and nonlinear systems and, therefore, very suitable for the analysis of complex telemanipulators.

5.3.3 Transparent Telemanipulation as a Behavioral Control Problem

It follows from the given definitions that in order to evaluate the transparency of a telemanipulation system a model of the environment is required. For the reasons discussed in Sec. 5.1, the Hunt-Crossly model is adopted. Free motion of the slave can be interpreted as the interaction of the slave with a particular environment whose Hunt-Crossley model has all the coefficients equal to 0. The dynamic structure of the model of the environment, therefore, will be always the same and what will change, depending on the kind of environment, will be the parameters characterizing the model. Master and slave controllers have to be designed in such a way that a certain degree of transparency is guaranteed. This control problem can be formulated in the behavioral framework:

Behavioral Control Problem 1 *Given an environment with a certain port-behavior $\mathcal{B}_e \subset \mathcal{U}$ represented by $\Delta_v^{\mathcal{B}_e,e}$ and a bound $a \in \mathbb{R}^+$ representing the desired degree of transparency, find a controller such that the transparency deviation index $\varepsilon = \|\Delta_v^{\mathcal{B}_e,e}(pr^{(n)}w_H)\|$ is such that $\varepsilon(t) \leq a \quad \forall t \in \mathbb{R}$*

A possible guideline for the determination of the transparency bound is the communication delay. In fact the transmission delay between master and slave side deteriorates transparency and the transparency degree that can be obtained is bounded by the communication delay. Furthermore, another fundamental factor to take into account in the design of the controller is that stability of the overall telemanipulation system MUST be guaranteed in all working conditions.

A telemanipulation system has to interact with several kinds of environment and, therefore, to keep on guaranteeing the same degree of transparency, the controllers have to change. Using Hunt-Crossley model, the model structure for the contact is the same; contacts with different environments have the same model but different parameters. Thus, in order to keep the same degree of transparency when interacting with different environment, it is sufficient to change the parameters of the controllers designed to solve Behavioral control Problem 1 according with the Hunt-Crossley model of the new environment.

Remark 5.9. In case of free motion of the slave, the controllers parameters can be chosen in order to optimize tracking performances according with any desired control technique (e.g. optimal control).

Assume that the controller port behavior is described by $\Delta_v^{\mathcal{B}_c,c}$, where c is the set of parameters characterizing the controller, and that a generic environment, using Hunt-Crossley model for the contact, is described by the differential structure $\Delta_v^{\mathcal{B}_e,e}$, where e is the set of parameters characterizing the environment. The generic problem to solve to have teleoperators with a certain degree of transparency when interacting with several environments is:

Behavioral Control Problem 2 *Given $a \in \mathbb{R}^+$ and m kind of environments that can be described by the Hunt-Crossley model and that are characterized by the following sets of respective parameters*

$$\{e_1, \dots, e_m\} \tag{5.15}$$

find m sets of controller parameters:

$$\{c_1, \dots, c_m\} \tag{5.16}$$

such that $\varepsilon = \|\Delta_v^{\mathcal{E},e_i}(pr^{(n)}(w_H(t),t)\|$ satisfies transparency requirements, that is

$$\varepsilon(t) \leq a \quad \forall t \in T \quad i = 1, \dots, m \tag{5.17}$$

Let i and j two kind of remote environments. When the teleoperator stops interacting with the environment i and start interacting with the environment j the parameters of the controller have to change form the set c_i to the set c_j in order to preserve the transparency degree. This operation of parameters adaptation has to be carefully addressed in order to avoid any unstable behavior in the transient. This behavioral control problem expresses the intuitive fact that to preserve a certain degree of transparency when interacting with different environments, it is necessary to change controller. Relying on the fact that contact with different environments can be modeled with the same model (Hunt-Crossley) characterized by different parameters, the change of controller is reduced to a parameter adaptation.

5.4 Transparency in Port-Hamiltonian Based Telemanipulation

The aim of this section is to apply the concepts developed so far to the port-Hamiltonian based intrinsically passive bilateral telemanipulation scheme discussed in Sec. 4.2. The considered telemanipulation scheme is reported in Fig. 4.5. Master and slave are represented as port-Hamiltonian systems and

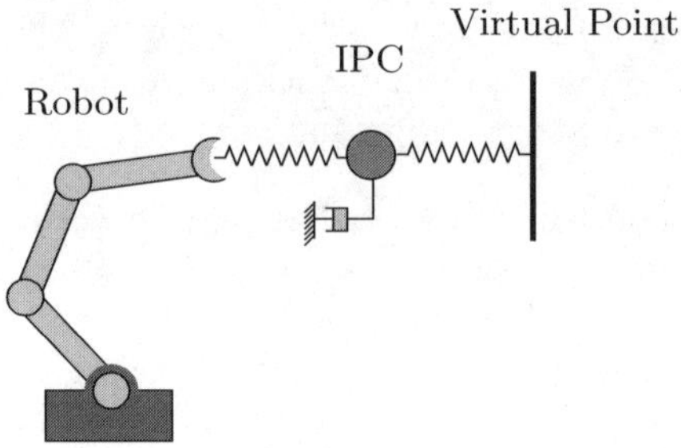

Fig. 5.2. The role of IPC.

the controllers are IPCs, which can be modeled as port-Hamiltonian systems and whose characterizing parameters have a direct physical meaning. The communication channel is implemented by means of scattering theory in order to have an intrinsically passive behavior independently of any communication delay.

Focus on the master side. The port-Hamiltonian controller can be interpreted as an interconnection of energy storing and of energy dissipating elements which connect the end-effector of the master with a *virtual point* in correspondence of which scattering transformation is performed as shown in Fig. 5.2 for an anthropomorphic robot. The virtual point represents the image of what happens at the slave side reported, by means of the communication channel, at the master side. The feeling the operator at master side gets of the interaction is that of the interaction between the master end-effector and the virtual point which is determined by the physical structure of the IPC as noted in [288]. Furthermore, also the impedance of the slave controller is perceived by the user as a part of the remote side and therefore its structure should be considered as well. Transparency can be deteriorated in two ways:

1. The structure of the IPCs does not allow to reproduce the port behavior of the remote environment.
2. The image of the slave side transmitted to the master side is deteriorated by the communication channel that introduces further spurious effects beyond the instantaneous power exchange.

5.4.1 Tuning the IPC

The Hunt-Crossley model can be physically interpreted as a mechanical parallel of a non linear spring and a non linear damper, that is the interconnection between an energy storing and an energy dissipating element. This interconnection can be modeled as a port-Hamiltonian system and it is an intrinsically passive system. Since the feeling the operator gets depends on the physical structure of the controllers and since the contact can be modeled by means of Hunt-Crossley model, which has a well determined physical structure, in order to improve the transparency of the telemanipulation scheme, port-Hamiltonian controllers with a Hunt-Crossley structure should be used. In this way the operator feels a compliant contact whose kind of behavior mimics that of the environment the slave is interacting with. Either by direct knowledge or by means of some identification algorithms proposed in the literature (e.g. [70]), the parameters of the Hunt-Crossley model of the remote environment can be estimated and used to modulate the controllers. Once the parameters of the IPCs have been modulated, it will represent a model of the remote environment and the operator will have a much more realistic feedback of the interaction of the slave with the virtual environment. Summarizing, the transparency-based IPCs design is composed of two steps:

1. **Use an IPC with Hunt-Crossley structure.** This addresses the Behavioral control Problem 1. The sensation that the operator feels depends on the kind of impedance controller used and therefore the port-Hamiltonian controller has to reflect the physical structure of the remote environment.
2. **Vary IPC parameters when interacting with different environments.** This addresses the Behavioral control Problem 2. Assuming that the remote environments can be represented by the Hunt-Crossley model, it is sufficient to change the parameters of the controller when interacting with different environments in order to guarantee a given degree of transparency.

Remark 5.10. Passivity of the controllers has to be always guaranteed. In Sec. 5.5 it will be shown how to change parameters of a port-Hamiltonian system while preserving passivity.

Remark 5.11. In order to avoid the phenomenon of wave reflection arising with scattering based communication channels, the line damper has to be embedded in the port-Hamiltonian controllers. This extra dynamics has to be taken into account when performing transparency analysis of the system.

5.4.2 Transparency Analysis of a Scattering Based Packet Switching Communication Channel

In the port-Hamiltonian based telemanipulation systems over packet switching networks, see Sec. 4.4, master and slave sides exchange energy through the communication channel to which they are connected by means of the discrete power ports $(e_m(k), f_m(k))$ and $(e_s(k), f_s(k))$ respectively. In case of negligible communication delay, the interconnection between local and remote sides can be made through the so called *common effort interconnection*, which is given by

$$\begin{cases} e_m(k) = e_s(k) \\ f_s(k) = -f_m(k) \end{cases} \tag{5.18}$$

and which implies that

$$P_s(k) = e_s^T(k)f_s(k) = -e_m^T(k)f_m(k) = P_m(k) \tag{5.19}$$

namely that the energy provided at the slave side at the time k is equal to that extracted from the master side at the same instant. Thus, in this ideal case, master and slave would directly exchange energy without any delay and without any loss of information. In case of non negligible communication delay, the strategy reported in Eq.(5.18) cannot be used since it would destabilize the overall telemanipulation system, as proven in Proposition 4.2.

The power ports $(e_m(k), f_m(k))$ and $e_s(k), f_s(k))$ can be represented as pairs of discrete scattering variables $(s_m^+(k), s_m^-(k))$ and $(s_s^+(k), s_s^-(k))$ respectively, where:

$$\begin{cases} s_*^+(k) = \frac{1}{\sqrt{2}}N^{-1}(e_*(k) + Zf_*(k)) \\ \\ s_*^-(k) = \frac{1}{\sqrt{2}}N^{-1}(e_*(k) - Zf_*(k)) \end{cases} \quad * = m, s \tag{5.20}$$

Combining Eq.(5.20) with Eq.(5.18), the common effort interconnection can be equivalently restated in terms of scattering variables as:

$$\begin{cases} s_m^+(k) = s_s^-(k) \\ \\ s_s^+(k) = s_m^-(k) \end{cases} \tag{5.21}$$

As proven in Sec. 4.4 this communication strategy can be safely used also in case of non negligible delay leading to the following interconnection:

$$\begin{cases} s_m^+(k) = s_s^-(k-\delta) \\ \\ s_s^+(k) = s_m^-(k-\delta) \end{cases} \tag{5.22}$$

where δ is the communication delay between master and slave sides and viceversa.

Remark 5.12. The results of this subsection can be easily generalized in case the communication delays between master and slave sides and viceversa are different.

Thus scattering theory can be used to allow a non destabilizing exchange of energy between master and slave; nonetheless, both transmission delay and the scattering coding/decoding procedure, necessary to compute $f(k)$ and $s^+(k)$ from $e(k)$ and $s^-(k)$ as described in Sec. 4.2.4, introduce some deleterious effects on transparency of the communication channel.

Remark 5.13. In telemanipulation schemes where the communication channel has an admittance causality (effort in/flow out) both at master and slave sides, the absence of delay leads to causality problems, as reported in [299, 215]. The transparency analysis that follows can be adapted in order to include this case by simply considering an infinitesimal delay at the corresponding power interconnection as an ideal case.

The aim of this subsection is to provide a transparency analysis of the discrete scattering implementation of switched packets communication channel for port-Hamiltonian based telemanipulation described in Sec. 4.4 using the framework proposed in Sec. 5.3. This kind of communication channel can be also used for interconnecting master and slave sides of generic passivity based telerobotic systems and, therefore, the following considerations are useful also for non port-Hamiltonian based telemanipulation schemes.
The ideal port behavior $\mathcal{I}$ which guarantees a completely transparent interconnection between the master and the slave sides is given by the common effort interconnection. Using Eq.(5.18), the ideal port behavior can be represented as the kernel of the following operator:

$$\Delta^{\mathcal{I}}(k, f_m(k), f_s(k), e_m(k), f_m(k)) = \begin{pmatrix} f_m(k) + f_s(k) \\ \\ e_s(k) - e_m(k) \end{pmatrix} \tag{5.23}$$

where the port outcome reduces to

$$(f_m(k), f_s(k), e_m(k), e_s(k))$$

since the interconnection is not characterized by a configuration. Notice that this is a degenerate case of port behavior since it is simply represented by an algebraic relation between the power variables. It is possible to evaluate the transparency of the transmission line as the behavioral deviation from the ideal behavior $\mathcal{I}$ when the communication delay is present. Using Eq.(5.20) in Eq.(5.22), by lengthy but straightforward computations, the communication strategy can be equivalently restated in terms of efforts and flows as:

$$\begin{cases} f_s(k) = -f_m(k) + (f_m(k) - f_m(k-\delta)) + Z^{-1}(e_m(k-\delta) - e_s(k)) \\ \\ e_m(k) = e_s(k) - (e_s(k) - e_s(k-\delta)) - Z(f_m(k) + f_s(k-\delta)) \end{cases} \tag{5.24}$$

Thus, using Eq.(5.24) in Eq.(5.23) and applying Def. 5.8, the transparency deviation index evaluated for the delayed scattering based communication channel is given by:

$$\varepsilon = \| \begin{pmatrix} (f_m(k) - f_m(k-\delta)) + Z^{-1}(e_m(k-\delta) - e_s(k)) \\ (e_s(k) - e_s(k-\delta)) + Z(f_m(k) + f_s(k-\delta)) \end{pmatrix} \| \leq$$

$$\underbrace{\| \begin{pmatrix} (f_m(k) - f_m(k-\delta)) \\ (e_s(k) - e_s(k-\delta)) \end{pmatrix} \|}_{\varepsilon_1} + \underbrace{\| \begin{pmatrix} Z^{-1}(e_m(k-\delta) - e_s(k)) \\ Z(f_m(k) + f_s(k-\delta)) \end{pmatrix} \|}_{\varepsilon_2} \tag{5.25}$$

The transparency deviation index is bounded by a sum of sub-indexes. Each of these indexes represents the effect on transparency of a particular phenomenon occurring in the implementation of the transmission. The term ε_1 describes

the effect on transparency of the communication delay and it does not depend on the scattering implementation of the interconnection. The term ε_2, instead, depends on the scattering based nature of the channel and it derives from the coding/decoding procedure that is used for going from scattering variables to power variables ad viceversa.
If we used the simpler communication strategy

$$\begin{cases} f_s(k) = -f_m(k-\delta) \\ e_m(k) = e_s(k-\delta) \end{cases} \tag{5.26}$$

the transparency deviation index would exhibit only the term ε_1 and, therefore, the only factor affecting transparency would be the communication delay. Thus, why the communication strategy reported in Eq.(5.26) shouldn't be used instead of that reported in Eq.(5.22)? The answer depends on the use of the transmission line. If the line has to be used simply to transmit two signals from one side to another and viceversa, the communication strategy reported in Eq.(5.26) gives better performances than the one reported in Eq.(5.22). On the other hand if, as in telemanipulation, the channel is used to interconnect dynamical systems it is necessary to keep into account the dynamic evolution of the overall system. The communication strategy in Eq.(5.26) leads to a non passive interconnection and to a possibly unstable system which is unacceptable. The communication strategy in Eq.(5.22) leads to a lossless interconnection and to a stable behavior of the overall system.

Thus, ε_2 represents the price we have to pay in terms of transparency of the communication channel in order to achieve a lossless interconnection and proves that transparency and passivity (together with robustness with respect to time delay and intrinsic stability) are conflicting targets as informally stated in [155] for linear telemanipulators.
During the communication over an packet switching networks, it is possible that some packets get lost. Using the RNP strategy proposed in Sec. 4.4 for preserving passivity of the communication channel in case of loss of packets transmission lines, when a packet is not received, it is replaced with a null packet. Thus, let $\mathcal{L}_{ms}$ and $\mathcal{L}_{sm}$ be the set of instants at which a packet is not received in the communication between master and slave and viceversa respectively. The transmission line is described by:

$$\begin{cases} s_s^+(k) = (1-\alpha)s_m^-(k-\delta) \\ s_m^+(k) = (1-\beta)s_s^-(k-\delta) \end{cases} \tag{5.27}$$

where

$$\alpha = \begin{cases} 1 \; k \in \mathcal{L}_{ms} \\ 0 \; k \notin \mathcal{L}_{ms} \end{cases} \qquad \beta = \begin{cases} 1 \; k \in \mathcal{L}_{sm} \\ 0 \; k \notin \mathcal{L}_{sm} \end{cases} \tag{5.28}$$

Once again, using Eq.(5.20) in Eq.(5.27), by straightforward calculations , it is possible to restate the communication strategy in terms of power variables as

$$\begin{cases} f_s(k) = -f_m(k) + (f_m(k) - f_m(k-\delta)) + \\ +Z^{-1}(e_m(k-\delta) - e_s(k)) - \alpha Z^{-1} s_m^-(k-\delta) \\ \\ e_m(k) = e_s(k) - (e_s(k) - e_s(k-\delta)) - \\ -Z(f_m(k) + f_s(k-\delta)) - \beta s_s^-(k-\delta) \end{cases} \tag{5.29}$$

whence, using Eq.(5.23), it is possible to calculate the transparency deviation index which results

$$\varepsilon \leq \varepsilon_1 + \varepsilon_2 + \underbrace{\| \begin{pmatrix} \alpha Z^{-1} s_m^-(k-\delta) \\ \beta s_s^-(k-\delta) \end{pmatrix} \|}_{\varepsilon_3} \tag{5.30}$$

Thus, in case of packets loss, transparency decreases and the contribution to transparency deviation is proportional to the norm of the lost energy packets. This is again linked to the fact that transparency and passivity are conflicting targets. In fact, when packets are lost and the RNP strategy is used, it means that their power content is dissipated; thus, packet loss introduces dissipation in the communication channel rendering it a strictly passive instead of a lossless system. This increase of passivity leads to a decrease of transparency and the term ε_3 is as more significant as greater is the power associated to the lost packets. In other words the more the communication channel gets passive the less it gets transparent.
Several strategies have been developed to recover the packets lost in communication. In case a lost packet is replaced by a packet obtained by interpolation, the transmission is described by

$$\begin{cases} s_s^+(k) = (1-\alpha) s_m^-(k-\delta) + \alpha s_{mI}^-(k-\delta) \\ \\ s_m^+(k) = (1-\beta) s_s^-(k-\delta) + \beta s_{sI}^-(k-\delta) \end{cases} \tag{5.31}$$

where s_{mI}^- and s_{sI}^+ are the interpolated packets that replace the master and slave lost packets respectively. In this case, following the same procedure used to get Eq.(5.30), we have that the transparency deviation index is given by:

$$\varepsilon \leq \varepsilon_1 + \varepsilon_2 + \underbrace{\| \begin{pmatrix} \alpha Z^{-1}(s_m^-(k-\delta) - s_{mI}^-(k-\delta)) \\ \beta(s_s^-(k-\delta) - s_{sI}^-(k-\delta)) \end{pmatrix} \|}_{\varepsilon_3} \tag{5.32}$$

Now the term related to packets loss depends on the error introduced by the interpolation process.

In Sec. 4.5 a passivity preserving interpolation technique, which replaces a sequence of lost packets with an opportunely weighted linear interpolation of the received packets, has been proposed.

This algorithm requires a receiving and a transmitting buffer in order to replace lost packets with the interpolated ones and, therefore, it leads to an increase of the communication delay and, in general, to a possible increase of the terms ε_1 and ε_2. Thus the interpolation reduces the term relative to the packet loss in the transparency deviation index but the price to pay is a possible increase of the terms relative to the delayed communication. Therefore, before enabling any interpolation algorithm, it is necessary to check if the beneficial effect introduced by the interpolation is not overwhelmed by the effect introduced by the increase of delay. This can be done by performing a worst case analysis (e.g. through the Monte Carlo method) on the value of the signal exchanged and on their variation rate; in fact,loosely speaking, the faster are the dynamics of master and slave sides the more an increase in communication delay deteriorates transparency of the overall system and thus, in this case, the interpolation should be disabled in order to keep the delay as small as possible.

Suppose now that the communication delay is variable as often happens when using switching packet networks. The communication strategy, in case of variable delay, becomes

$$\begin{cases} s_s^+(k) = s_m^-(k - \delta + n(k)) \\ s_m^+(k) = s_s^-(k - \delta + n(k)) \end{cases} \tag{5.33}$$

where $n(k) \in \mathbb{Z}$ represents the variability of the delay; we suppose that the delay has the same variability both in the communication between master and slave and viceversa in order to keep the notation simple in the computations. The results obtained can be easily extended to the general case. We suppose that the variable delay is an undesired effect and that the communication channel is supposed to be characterized by a constant delay δ in both directions. Suppose that the delay is increasing, namely that $n(k+1) > n(k)$ and that the packets $s_s^+(k) = s_m^-(k-\delta+n(k))$ and $s_m^+(k) = s_s^-(k-\delta+n(k))$ have just been received; we then have:

$$\begin{cases} s_s^+(k+i) = 0 \quad i = 1, \ldots, [n(k+1) - n(k) - 1] \\ s_s^+(k + (n(k+1) - n(k))) = s_m^-(k - \delta + n(k) + 1) \end{cases}$$
$$\begin{cases} s_m^+(k+i) = 0 \quad i = 1, \ldots, [n(k+1) - n(k) - 1] \\ s_m^+(k + (n(k+1) - n(k))) = s_s^-(k - \delta + n(k) + 1) \end{cases} \tag{5.34}$$

where the RNP strategy is used and, when a packet is late, it is replaced with 0 in order to preserve passivity. Assuming that there is no packet loss to keep the notation simple and proceeding in the same way as done for computing the previous sub-indexes, we have that:

$$\varepsilon \leq \varepsilon_1 + \varepsilon_2 + \varepsilon_4 \tag{5.35}$$

where

$$\varepsilon_4(k+i) =$$

$$\begin{cases} \left\| \begin{pmatrix} Z^{-1} s_m^-(k-\delta+n(k)+i) \\ s_s^-(k-\delta+n(k)+i) \end{pmatrix} \right\| \\ \qquad\qquad i = 1, \ldots, [n(k+1) - n(k) - 1] \\ \left\| \begin{pmatrix} Z^{-1}(s_m^-(k-\delta+n(k)+1) - s_m^-(k-\delta+n(k)+n(k+1)-n(k)) \\ (s_s^-(k-\delta+n(k)+1) - s_s^-(k-\delta+n(k)+n(k+1)-n(k)) \end{pmatrix} \right\| \\ \qquad\qquad i = [n(k+1) - n(k)] \end{cases} \tag{5.36}$$

The effect on transparency of an increase of delay is twofold: for $n(k+1) - n(k) - 1$ samples it brings the same effect brought by a packet loss since, because of the increasing delay, the expected packet is not delivered on time and it is replaced by 0. Furthermore there is a second effect due to the fact that when finally the packet is delivered it is not the packet that is expected at that instant. Unlike for the case of packets loss, it is not possible to replace the "holes" due to the increase of delay by interpolation since finally the transmitted packets arrive. If we filled the holes with new packets we would introduce extra energy into the system leading to an active communication channel. Thus while preserving passivity is not a constraint for the recovery process for lost packets it is a constraint for the compensation of the effects due to variable delay. Suppose now that the communication delay is decreasing, namely that $n(k) > n(k+1)$. In order to avoid very long delays (that would take place in case of retransmission techniques were adopted) that would make a telemanipulation system quite useless, and to avoid wave distortion (due to the fact that packets does not arrive in the correct order) that would tremendously decrease the performances of the system, the most used strategy to handle decreasing delay is the Use Freshest Sample (UFS) strategy. A time stamp is attached to each transmitted packet and if a packet older than the last received packet arrives, it is simply discarded. Thus the effect of a decrease of delay on the transparency is the same as that given by a packet loss.

Summarizing, the framework for the analysis of transparency introduced in Sec. 5.3 allowed analyzing the scattering based switching packets communication strategy used for the interconnection of master and slave sides in port-Hamiltonian based telemanipulation. It has been possible to recognize various factors affecting transparency and to formally prove that transparency and passivity are conflicting targets. The transparency deviation index of the communication channel is bounded by the sum of four terms: ε_1 that takes into account the communication delay, ε_2 that considers the scattering coding/decoding procedure. The possible decrease in transparency due to the scattering based implementation is the price to pay in order to achieve a lossless transmission. The term ε_3 encodes the effect of packets loss and it can be optimally minimized in a passive way by replacing the lost packets by interpolated packets. Finally, the term ε_4 encodes the effects due to the variability of the time delay; these are the most critical effects since they cannot be compensated without affecting the passivity of the communication channel. In conclusion, in order to maximize the transparency of the interconnection between master and slave sides, it is necessary to choose a communication channel that, even if characterized by quite a big loss rate, has a little variable delay. In this way the unrecoverable transparency deviation due to variable delay is minimized and it is possible to passively reduce the effects of packet loss through the interpolation algorithm proposed in Sec. 4.5

5.4.3 Simulations

Consider a simple one degree of freedom telemanipulation scheme where master and slave robots are simple masses, the environment the system is interacting with is a parallel of a spring and a damper with known damping factor and the IPCs are implemented as the equivalent of a parallel between a spring and a damper. The communication channel is implemented by means of scattering theory. Simulations are performed in the following situation. Master and slave have a mass equal to 1 Kg and the characteristic impedance of the scattering based communication channel is $Z = 1\ Nsec/m$ and the telemanipulation system is interacting with an environment made up with a parallel of a spring with stiffness $k_e = 100\ N/m$ and a damper whose damping coefficient is $b_e = 10\ Nsec/m$. The human operator pushes the master with a force with sinusoidal profile, as if he/she were probing the environment. The communication delay is $\delta = 0.5\ sec$.

The goal of the simulation is to show the effect of a proper tuning of the IPCs. Two sets of IPCs parameters have been considered

1. **Set 1:** $b = 1\ Nsec/m$ and $k = 1\ N/m$
2. **Set 2:** $b = 10\ Nsec/m$ and $k = 100\ N/m$

The results have been plotted in Fig. 5.3. It can be seen that when the parameters of the IPC match the parameters of the environment, the transparency index gets much lower.
Suppose now that master and slave sides are interconnected by means of a packet switched communication channel with nominal delay $\delta = 0.2\ sec$ and that the operator applies an impulsive force on the master. The goal of the next set of simulations is to show the effect of the phenomena characterizing the packet switching networks on the transparency of the communication channel. In the first simulation we implemented a packet loss, with loss rate of 50% in both senses of communication. The terms relative to transparency deviation and the position of master and slave are reported in Fig. 5.4.

We can see that all the transparency sub-indexes tends to zero. This is due to the fact that after a certain transient the system stops and, therefore, zero efforts and zero flows are exchanged along the communication channel that, therefore, appears completely transparent. During the motion the transparency deviation terms are different form zero meaning that the communication

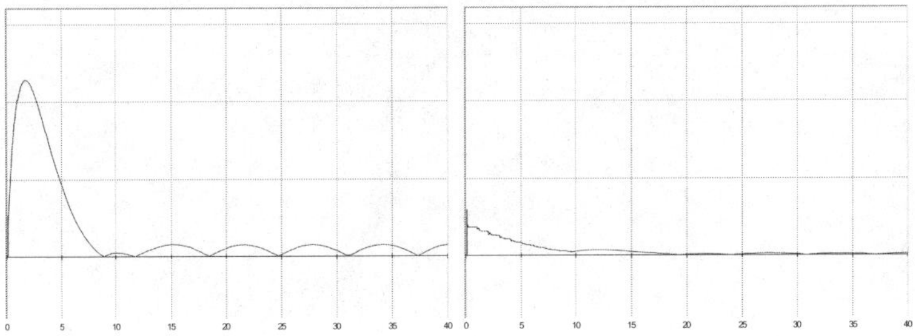

(a) Set 1 (b) Set 2

Fig. 5.3. $\varepsilon(t)$: effect of controllers parameters

channel is not transparent; in particular, ε_3 exhibits peaks that correspond to the packets lost in the communication. A non transparent communication channel leads to a non transparent telemanipulation system as can be noticed by the positions of master and slave that are quite different.

In Fig. 5.5 the behavior of the communication channel is reported when the interpolation algorithm proposed in Sec. 4.5 is enabled. We can see that the term ε_1 slightly increases because of the increase of delay induced by the receiving buffer. Nevertheless, this slight increase is greatly repaid by the decrease of the term ε_3 because of the optimal recovery of the sequences of lost packets. The overall decrease of transparency deviation of the communication channel can be observed in an increase of performances in the positioning task; in fact, now, the position of the slave is closer to that of the master. In the last simulation we implement a variable communication delay where the UFS strategy is adopted. The simulation results are reported in Fig. 5.6. We can see that both ε_1 and ε_2 increase because of the variability of the delay. Furthermore the term ε_4, which encodes the effect of variable delay, is the most significant transparency deviation term. The effect of variable delay is the most deleterious since no action can be taken to compensate it without affecting the passivity of the communication. The performances of the telemanipulation system are quite bad, as it can be noticed by looking at the positions of master and slave, but, nevertheless, the system keeps on being stable because the communication channel keeps on being passive as proven in Sec. 4.4.

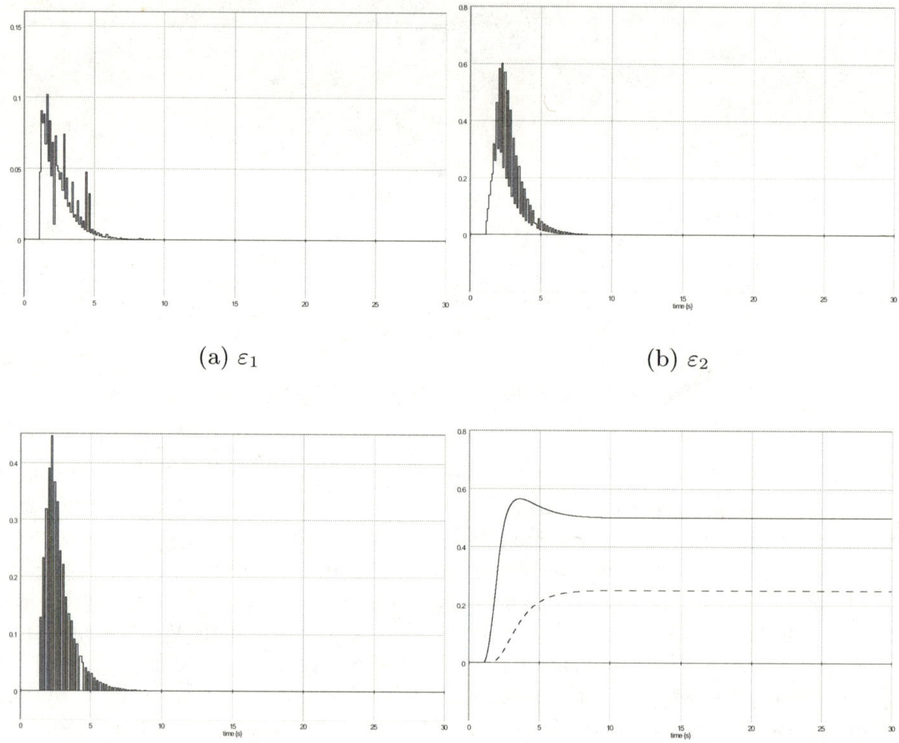

(a) ε_1 (b) ε_2

(c) ε_3 (d) Position of master (solid) and slave (dashed)

Fig. 5.4. Effect on transparency of loss of packets in the communication

5.5 A Passivity Preserving Tuning of Port-Hamiltonian Systems

A port-Hamiltonian system can be thought as a modulated interconnection of energy storing and energy dissipating elements. From this point of view parameters can appear in:

- *Energy storing elements*: these parameters have a direct physical meaning (e.g. stiffness in springs) and play a role in the energy storage.
- *Energy dissipating elements*: these parameters have a clear physical meaning (e.g. damping coefficient in Colombian friction) and play a role in the energy dissipation.

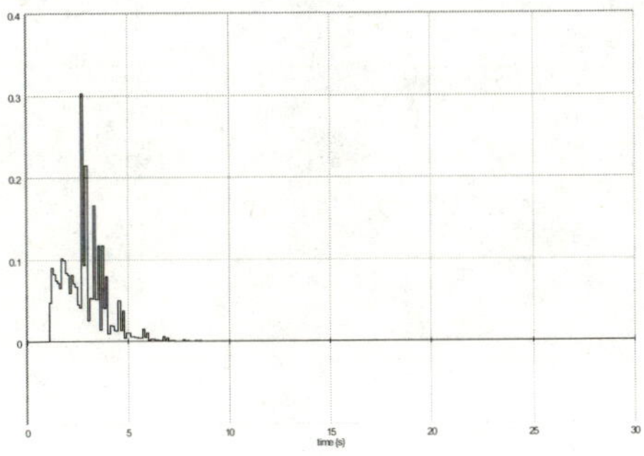

(a) ε_1

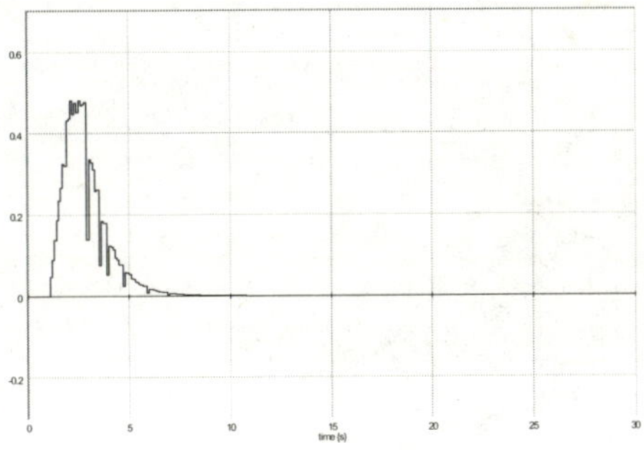

(b) ε_2

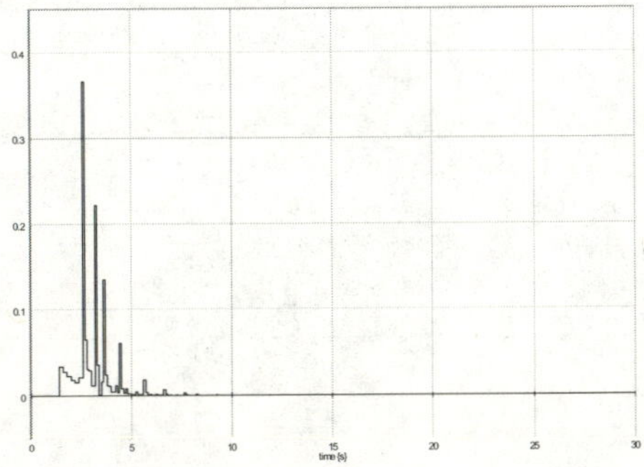

(c) ε_3

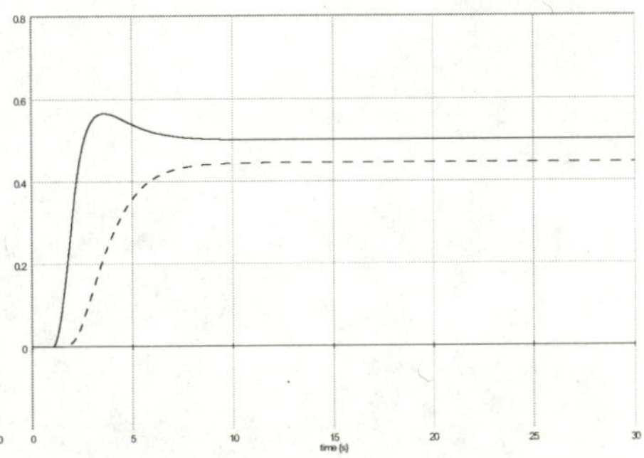

(d) Position of master (solid) and slave (dashed)

Fig. 5.5. Effect on transparency of loss of packets in the communication when interpolation is enabled

- *Interconnection*: these parameters play a role in the interconnection modulation and, therefore, in the way energy flows among all the elements (e.g. electro-mechanical coupling in DC motors).

When port-Hamiltonian systems are used as controllers (e.g. in telemanipulation) it can be very useful to change physical parameters to improve performances (e.g. to improve transparency in bilateral telemanipulation); however, parameters play an explicit role in the energetic behavior of the system and a regardless parametric variation can introduce extra energy.

Example 5.14. Consider a linear spring with stiffness $k = 1\ N/m$ and with length fixed at $x = 1\ m$. The energy stored in the spring is given by the well known formula:

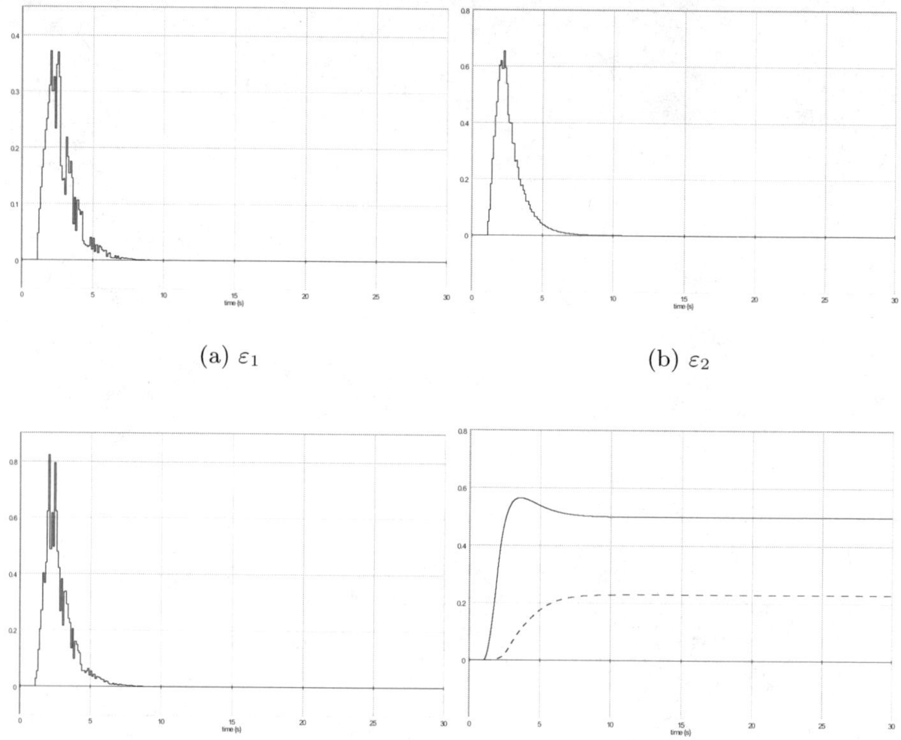

(a) ε_1 (b) ε_2

(c) ε_4 (d) Position of master (solid) and slave (dashed)

Fig. 5.6. Effect on transparency of variable delay in the communication

$$E(t) = \frac{1}{2}kx^2(t) = 0.5\ J$$

Assume now to change k from 1 N/m to 2 N/m while keeping the elongation fixed. The energy stored in the spring is now:

$$E(t) = \frac{1}{2}kx^2(t) = 1\ J$$

It can be clearly seen that 0.5 J have been produced by changing the stiffness.

As shown in Example 5.14, parameters variation must be carefully addressed in order to avoid extra energy production and, therefore, loss of passivity.

The main idea for changing parameters while preserving passivity is to consider parameters as states instead of constants and to use part of the energy flowing through the system to change their value. In this way no extra energy is introduced and passivity is preserved. We will start from a fixed parameters port-Hamiltonian system and we will obtain what we call *extended port-Hamiltonian system*: we will extend the four basic ingredients $(\mathcal{X}, \mathcal{V}, \mathcal{D}, \mathcal{H})$ necessary to define a fixed parameters port-Hamiltonian system, namely the state manifold, the Dirac structure, the Hamiltonian function and the spaces of flows and effort variables (see Sec. 1.4), to obtain a variable parameters port-Hamiltonian system where passivity is preserved.

5.5.1 Parameters Associated to Energy Storage

In port-Hamiltonian systems energy storage is represented by the state $x \in \mathcal{X}$ AND fixed physical parameters (e.g. stiffnesses). The storage of a certain amount of energy causes a variation in the state of the system. Modeling each physical parameter as a state instead of a constant value, we obtain that in front of some energy storage physical parameters can be changed as well. In the latter case it is necessary to find a new Dirac structure that models the flow of energy which is stored in the "new" states. We call this new Dirac structure an *extended* Dirac structure; furthermore, the part of the state variables that represents physical parameters is called *parametric part* of the state, while the part of the state which does not represent physical parameters is called *non parametric part* of the state.

Remark 5.15. We do NOT want to replace the Dirac structure of the fixed parameters port-Hamiltonian system with a brand new one since we want the internal energetic interconnection, and, therefore, the kind of behavior of the system, to be preserved. We rather speak of *extension* since we just want to add some new interconnections which allow to deviate a part of the internal energy towards states representing physical parameters enabling, thus, their variation.

Assume to have a fixed parameters port-Hamiltonian system where $\mathcal{X}$ is an n-dimensional manifold representing the state space, $\mathcal{V}$ is a m-dimensional vector space that, with its dual m-dimensional vector space $\mathcal{V}^*$, represents the power port structure, $\mathcal{D}$ is the Dirac structure modeling the internal interconnection and $\mathcal{H} : \mathcal{X} \to \mathbb{R}$ is the Hamiltonian function representing the energy of the system. In order to allow passive parametric variation we need to model parameters as additional states. Let $\mathcal{K}$ be a k-dimensional manifold representing the parameters space. In the extended formulation, the $n + k$-dimensional manifold $\mathcal{X}_e = \mathcal{X} \times \mathcal{K}$ represents the extended state space; once local coordinates have been fixed , an element $x_e \in \mathcal{X}_e$ can be represented by the $n + k$-tuple $x_e = (x, k)$ where x and k represent the non parametric and the parametric part of the state respectively. The following differential structures are well defined:

- $T_{x_e}\mathcal{X}_e = T_x\mathcal{X} \times T_k\mathcal{K}$: tangent space at the point $x_e \in \mathcal{X}_e$.
- $T\mathcal{X}_e = \cup_{x_e \in \mathcal{X}_e} T_{x_e}\mathcal{X}_e$: tangent bundle of $\mathcal{X}_e$.

- $T^*_{x_e}\mathcal{X}_e = T^*_x\mathcal{X} \times T^*_k\mathcal{K}$: cotangent space at the point $x_e \in \mathcal{X}_e$.
- $T^*\mathcal{X}_e = \cup_{x_e \in \mathcal{X}_e} T^*_{x_e}\mathcal{X}_e$: cotangent bundle of $\mathcal{X}_e$.

Hence, there exists an intrinsic duality product on $\mathcal{X}_e$, denoted by $\langle , \rangle$, being a bilinear map from $T\mathcal{X}_e \times T^*\mathcal{X}_e$ to $\mathbb{R}$ defined as:

$$\langle v_e^*, v_e \rangle = v_e^*(v_e) \qquad v_e \in T_{x_e}\mathcal{X}_e, v_e^* \in T^*_{x_e}\mathcal{X}_e, x_e \in \mathcal{X}_e \tag{5.37}$$

Since we want to use part of the *internal* energy to change parameters, there is no need to change the power port structure. The power port structure, therefore, is still represented by the same m-dimensional vector space $\mathcal{V}$ and its dual $\mathcal{V}^*$ and their intrinsically defined duality product. We can, therefore, consider the following vector bundle (see Sec. A.1)over the extended space manifold $\mathcal{X}_e$.

$$\mathcal{Q}_e = T\mathcal{X}_e \oplus T^*\mathcal{X}_e \oplus \mathcal{V} \oplus \mathcal{V}^* \tag{5.38}$$

The fibers of the bundle $\mathcal{Q}_e$ define the space of extended vectors, extended covectors, flows and efforts.

$$\mathcal{Q}(x_e) = T_{x_e}\mathcal{X}_e \times T^*_{x_e}\mathcal{X}_e \times \mathcal{V} \times \mathcal{V}^* \tag{5.39}$$

A Dirac structure can, thus, be defined on the extended state space manifold as a vector sub-bundle $\mathcal{D}_e \subset \mathcal{Q}_e$ using Def. 1.21 after having introduced +-pairing on $\mathcal{Q}_e$ analogously to what done in Sec. 1.4.
It can be easily shown that:

$$\dim \mathcal{D}(x_e) = \dim \mathcal{X}_e + \dim \mathcal{V} = \dim \mathcal{X} + \dim \mathcal{K} + \dim \mathcal{V} = n + k + m \tag{5.40}$$

Furthermore, being $\mathcal{D}_e$ a Dirac structure we have that:

$$\langle v_e, v_e^* \rangle + \langle e, f \rangle = \frac{1}{2}\langle (v_e, v_e^*, f, e), (v_e, v_e^*, f, e) \rangle_+ = 0 \quad \forall (v_e, v_e^*, f, e) \in \mathcal{D}_e(x_e) \tag{5.41}$$

Both in fixed and variable parameters port-Hamiltonian systems the energy is a function is the same. The difference is that in the first case physical parameters are constant while in the second case parameters can change. Thus, the Hamiltonian function for the extended case has the same structure of the Hamiltonian function for the fixed parameters case.

Definition 5.16. *Consider the extended state space $\mathcal{X}_e$, the space of flow variables $\mathcal{V}$ and, dually, the space of effort variables $\mathcal{V}^*$, the extended Dirac structure $\mathcal{D}_e$ and the Hamiltonian function $\mathcal{H}$. Then the extended port-Hamiltonian system corresponding to the 4-tuple $(\mathcal{X}_e, \mathcal{V}, \mathcal{D}_e, \mathcal{H})$ is defined by setting:*

$$v_e = -\dot{x}_e = -\begin{pmatrix} \dot{x} \\ \dot{k} \end{pmatrix} \qquad v_e^* = \frac{\partial \mathcal{H}}{\partial x_e} = \begin{pmatrix} \frac{\partial \mathcal{H}}{\partial x} \\ \\ \frac{\partial \mathcal{H}}{\partial k} \end{pmatrix} \tag{5.42}$$

that is by the equations:

$$(-\dot{x}_e, \frac{\partial \mathcal{H}}{\partial x_e}, f, e) \in \mathcal{D}_e(x_e) \tag{5.43}$$

The property reported in Eq.(5.41) immediately yields the following power balance:

$$\dot{\mathcal{H}}(t) = \langle \frac{\partial \mathcal{H}}{\partial x_e}, \dot{x}_e \rangle = \langle \frac{\partial \mathcal{H}}{\partial x}, \dot{x} \rangle + \langle \frac{\partial \mathcal{H}}{\partial k}, \dot{k} \rangle = \langle e, f \rangle \tag{5.44}$$

that is an extended port-Hamiltonian system is *lossless.*
In the fixed parameters case (see Eq.(1.31)) the energy balance doesn't involve explicitly parameters , now (see Eq.(5.44)) part of the energy is stored in the parametric part of the state allowing thus variation of physical parameters. Deviating part of the energy to be stored towards parameters allows to vary parameters and to preserve losslessness of port-Hamiltonian systems.

Analogously to the fixed parameters case, we can consider dissipation by terminating some ports with energy absorbing elements. We have seen that it is possible to extend a fixed parameters port-Hamiltonian system by simply considering physical parameters as states. Considering the extended state manifold it is still possible to define a (extended) port-Hamiltonian system with the same energy function and the same power-port structure as the original system. Using the extended state space is possible to build Dirac structures which take into account the possibility of storing some energy in the parametric part of the state allowing, therefore, variations of physical parameters.

According to Remark 5.15 we would like to build a Dirac structure $\mathcal{D}_e$ on $\mathcal{X}_e$ which is an extension of the Dirac structure $\mathcal{D}$ of the fixed parameters system on $\mathcal{X}$. Until now we have shown that there exist Dirac structures on $\mathcal{X}_e$ that allow energy storage in the parametric part of the state. We will show, in a coordinate based approach for the sake of clearness, how to choose, among the extended Dirac structures, the ones which are extensions $\mathcal{D}$.

Remark 5.17. There is not, in principle, a unique way to design the extension of the Dirac structure. The choice of a particular extension depends on the task to be performed and on the kind of parametric variation needed.

Choosing local coordinates, it is possible to find a representation of extended Dirac structures, analogously to what done in Eq.(1.46), and consequently of extended port-Hamiltonian systems with dissipation, analogously to what done in Eq.(1.50). A representation of an extended port-Hamiltonian system with dissipation is given by:

$$\underbrace{\left(F_{se}(x_e)\; F_{pe}(x_e)\; F_{re}(x_e)\right)}_{\tilde{F}_e(x_e)} \begin{pmatrix} -\dot{x}_e \\ f \\ f_r \end{pmatrix} + \underbrace{\left(E_{se}(x_e)\; E_{pe}(x_e)\; E_{re}(x_e)\right)}_{\tilde{E}_e(x_e)} \begin{pmatrix} \frac{\partial \mathcal{H}}{\partial x_e} \\ e \\ e_r \end{pmatrix} = 0 \tag{5.45}$$

where $\dot{x}_e = (\dot{x}, \dot{k})$ and $\tilde{F}_e(x_e)$ and $\tilde{E}_e(x_e)$ are $(n+k+m+r) \times (n+k+m+r)$ matrices for which:

$$\begin{cases} \tilde{F}_e(x_e)\tilde{E}_e^T(x_e) + \tilde{E}_e(x_e)\tilde{F}_e^T(x_e) = 0 \\ \\ rank[\tilde{F}_e(x_e) : \tilde{E}_e(x_e)] = n+k+m+r \end{cases} \tag{5.46}$$

and $F_{se}(x_e), E_{se}(x_e)$ are a $(n+k) \times (n+k+m+r)$ matrices, $F_{pe}(x_e), E_{pe}(x_e)$ are $m \times (m+k+m+r)$ matrices and $F_{re}(x_e), E_{re}(x_e)$ are $r \times (n+k+m+r)$ matrices.

Assume to have a fixed parameters port-Hamiltonian system represented by the following matrices:

$$\tilde{F}(x) = \left(F_s(x)\; F_p(x)\; F_r(x)\right) \quad \tilde{E}(x) = \left(E_s(x)\; E_p(x)\; E_r(x)\right) \tag{5.47}$$

where

$$\begin{aligned} F_s(x) &= \begin{pmatrix} F_{ss}(x) \\ F_{sp}(x) \\ F_{sr}(x) \end{pmatrix} \; F_p(x) = \begin{pmatrix} F_{ps}(x) \\ F_{pp}(x) \\ F_{pr}(x) \end{pmatrix} \; F_r(x) = \begin{pmatrix} F_{rs}(x) \\ F_{rp}(x) \\ F_{rr}(x) \end{pmatrix} \\ E_s(x) &= \begin{pmatrix} E_{ss}(x) \\ E_{sp}(x) \\ E_{sr}(x) \end{pmatrix} \; E_p(x) = \begin{pmatrix} E_{ps}(x) \\ E_{pp}(x) \\ E_{pr}(x) \end{pmatrix} \; E_r(x) = \begin{pmatrix} E_{rs}(x) \\ E_{rp}(x) \\ E_{rr}(x) \end{pmatrix} \end{aligned} \tag{5.48}$$

Since $\tilde{F}(x)$ and $\tilde{E}(x)$ represent a port-Hamiltonian system we have that:

$$\tilde{F}(x)\tilde{E}^T(x) + \tilde{E}(x)\tilde{F}^T(x) = 0 \tag{5.49a}$$

$$rank[\tilde{F}(x) : \tilde{E}(x)] = n+m+r \tag{5.49b}$$

Proposition 5.18. *Fixed parameters port-Hamiltonian systems are a particular case of extended port-Hamiltonian systems.*

Proof. Consider the following matrices:

$$\begin{aligned} F_{se}(x_e) &= \begin{pmatrix} F_{ss}(x) & 0 \\ 0 & I \\ F_{sp}(x) & 0 \\ F_{sr}(x) & 0 \end{pmatrix} \; F_{pe}(x_e) = \begin{pmatrix} F_{ps}(x) \\ 0 \\ F_{pp}(x) \\ F_{pr}(x) \end{pmatrix} \; F_{re}(x_e) = \begin{pmatrix} F_{rs}(x) \\ 0 \\ F_{rp}(x) \\ F_{rr}(x) \end{pmatrix} \\ E_{se}(x_e) &= \begin{pmatrix} E_{ss}(x) & 0 \\ 0 & 0 \\ E_{sp}(x) & 0 \\ E_{sr}(x) & 0 \end{pmatrix} \; E_{pe}(x_e) = \begin{pmatrix} E_{ps}(x) \\ 0 \\ E_{pp}(x) \\ E_{pr}(x) \end{pmatrix} \; E_{re}(x_e) = \begin{pmatrix} E_{rs}(x) \\ 0 \\ E_{rp}(x) \\ E_{rr}(x) \end{pmatrix} \end{aligned} \tag{5.50}$$

and build matrices $\tilde{F}_e(x_e)$ and $\tilde{E}_e(x_e)$ representing an extended port-Hamiltonian system as:

$$\tilde{F}_e(x_e) = \big(F_{se}(x_e)\ F_{pe}(x_e)\ F_{re}(x_e)\big)\ \tilde{E}_e(x_e) = \big(E_{se}(x_e)\ E_{pe}(x_e)\ E_{re}(x_e)\big) \tag{5.51}$$

By straightforward calculations, using Eq.(5.51) with Eq.(5.45), we obtain the following relations.

$$\begin{cases} -F_{ss}(x)\dot{x} + F_{sp}(x)f + F_{sr}(x)f_r + E_{ss}\frac{\partial \mathcal{H}}{\partial x} + E_{sp}(x) + E_{sr}(x)e_r = 0 \\ -F_{ps}(x)\dot{x} + F_{pp}(x)f + F_{pr}(x)f_r + E_{ps}\frac{\partial \mathcal{H}}{\partial x} + E_{pp}(x) + E_{sp}(x)e_r = 0 \\ -F_{rs}(x)\dot{x} + F_{rp}(x)f + F_{rr}(x)f_r + E_{rs}\frac{\partial \mathcal{H}}{\partial x} + E_{rp}(x) + E_{rr}(x)e_r = 0 \end{cases} \tag{5.52a}$$

$$\dot{k} = 0 \tag{5.52b}$$

The dynamic relations between x, flows and effort represented in Eq.(5.52a) are exactly the same as the one represented by the fixed parameter port-Hamiltonian system (use Eq.(5.47) with Eq.(1.50)). The behaviors of fixed and extended port-Hamiltonian systems could be different if the parametric part of the extended case were not constant, but Eq.(5.52b) says that parameters are constant. We can, therefore, conclude that the extended port-Hamiltonian system given by Eq.(5.51) represents the fixed parameter port-Hamiltonian system given in Eq.(5.47).

We need to prove that $\tilde{E}_e(x_e)$ and $\tilde{F}_e(x_e)$ represent a Dirac structure, namely that Eq.(5.46) is satisfied. Since $\tilde{E}(x)$ and $\tilde{F}(x)$ represent a Dirac structure Eq.(5.49a) holds and by lengthy but straightforward calculations it follows that:

$$\tilde{E}_e(x_e)\tilde{F}_e^T(x_e) + \tilde{F}_e(x_e)\tilde{E}_e^T(x_e) = 0$$

Furthermore we have that:

$$rank[\tilde{E}_e(x_e) : \tilde{F}_e(x_e)] =$$

$$rank \begin{pmatrix} E_{ss}(x) & 0 & E_{ps}(x) & E_{rs}(x) & F_{ss}(x) & 0 & F_{ps}(x) & F_{rs}(x) \\ 0 & 0 & 0 & 0 & 0 & I & 0 & 0 \\ E_{sp}(x) & 0 & E_{pp}(x) & E_{rp}(x) & F_{sp}(x) & 0 & F_{pp}(x) & F_{rp}(x) \\ E_{sr}(x) & 0 & E_{pr}(x) & E_{rr}(x) & F_{sr}(x) & 0 & F_{pr}(x) & F_{rr}(x) \end{pmatrix}$$

Since $\tilde{E}(x)$ and $\tilde{F}(x)$ represent a Dirac structure, Eq.(5.49b) holds and, therefore, the sub-matrix:

$$S = \begin{pmatrix} E_{ss}(x) & 0 & E_{ps}(x) & E_{rs}(x) & F_{ss}(x) & 0 & F_{ps}(x) & F_{rs}(x) \\ E_{sp}(x) & 0 & E_{pp}(x) & E_{rp}(x) & F_{sp}(x) & 0 & F_{pp}(x) & F_{rp}(x) \\ E_{sr}(x) & 0 & E_{pr}(x) & E_{rr}(x) & F_{sr}(x) & 0 & F_{pr}(x) & F_{rr}(x) \end{pmatrix}$$

has rank $n + m + r$. The matrix $[\tilde{E}_e(x_e) : \tilde{F}_e(x_e)]$ is obtained by adding k independent rows to the sub-matrix S and, therefore, we can conclude that $rank[\tilde{E}_e(x_e) : \tilde{F}_e(x_e)] = n + k + m + r$.

The proposed extension is, therefore, well posed since fixed parameters port-Hamiltonian systems can be expressed as a particular case of extended port-Hamiltonian systems.

In order to extend a fixed parameters port-Hamiltonian system to add the capability of changing physical parameters, we need to extend the fixed parameters port-Hamiltonian representation given in Eq.(5.51). The main idea is to build the extended Dirac structure in such a way that the extended system allows a passive variation of physical parameters while preserving a physical behavior as close as possible to the one of the fixed parameters port-Hamiltonian system; loosely speaking, we want to preserve the same kind physical behavior and to be able change physical parameters that characterize it. A simple approach to solve the problem is to *add* some dynamical relations in order to allow to store some energy in the parametric part of the state leaving existing relations, determined by the original system Dirac structure, unchanged. In coordinate based terms, the idea is to replace the zeros in Eq.(5.47) by proper terms that allow parameters variation preserving the Dirac structure and, thus, passivity. Two meaningful choices can be done. A first one is:

$$\tilde{F}_e(x_e) = \begin{pmatrix} F_{ss}(x) & 0 & F_{ps}(x) & F_{rs}(x) \\ 0 & I & Z & 0 \\ F_{sp}(x) & 0 & F_{pp}(x) & F_{rp}(x) \\ F_{sr}(x) & 0 & F_{pr}(x) & F_{rr}(x) \end{pmatrix}$$
$$\tilde{E}_e(x_e) = \begin{pmatrix} E_{ss}(x) & 0 & E_{ps}(x) & E_{rs}(x) \\ 0 & 0 & 0 & 0 \\ E_{sp}(x) & -Z^T & E_{pp}(x) & E_{rp}(x) \\ E_{sr}(x) & 0 & E_{pr}(x) & E_{rr}(x) \end{pmatrix} \tag{5.53}$$

and a second one is:

$$\tilde{F}_e(x_e) = \begin{pmatrix} F_{ss}(x) & 0 & F_{ps}(x) & F_{rs}(x) \\ 0 & I & 0 & 0 \\ F_{sp}(x) & 0 & F_{pp}(x) & F_{rp}(x) \\ F_{sr}(x) & 0 & F_{pr}(x) & F_{rr}(x) \end{pmatrix}$$
$$\tilde{E}_e(x_e) = \begin{pmatrix} E_{ss}(x) & -F_{ss}(x)Z & E_{ps}(x) & E_{rs}(x) \\ Z^T & 0 & 0 & 0 \\ E_{sp}(x) & -F_{sp}(x)Z & E_{pp}(x) & E_{rp}(x) \\ E_{sr}(x) & -F_{sr}(x)Z & E_{pr}(x) & E_{rr}(x) \end{pmatrix} \tag{5.54}$$

In the choice made in Eq.(5.53), part of the energy that comes from the interconnection and that has to be stored in the non-parametric part of the state is intercepted and deviated towards the parametric part of the state. In the choice made in Eq.(5.54) the energy needed for parametric variation is taken from the energy already stored in the non parametric part of the state.

Remark 5.19. The presence of F_{ss}, F_{sp} and F_{sr} in Eq.(5.54) is due to the fact that the interaction between parametric and non parametric part of the extended state has to take into account the algebraic constraints in the non parametric part of the extended state, the non parametric part of the state and port-flows and the non parametric part of the state and the resistive flows.

Proposition 5.20. *Matrices $\tilde{E}_e(x_e)$ and $\tilde{F}_e(x_e)$ reported in Eq.(5.53) and Eq.(5.54) represent Dirac structures.*

Proof. The proof follows from direct verification of the conditions reported in Eq.(5.46).

We imposed an energetic interaction between the parametric and non parametric parts of the state. This "energy sharing" allows parameters to change either when there is energy storage or when there is an energy flow to store. Usually the matrix Z, regulating the variation of physical parameters, has the following form:

$$Z = \sigma Z_1 \quad \sigma \in \{0, 1\} \tag{5.55}$$

The scalar σ plays the role of a switch that enables/disables the parametric variation. In several application (e.g. telemanipulation) when a certain set of physical parameters is reached, it has to be kept constant despite of any internal energy flows. Thus, when it is necessary to change the parameters, σ is set equal to 1 and when a set of parameters has to e kept constant σ is set t 0.

Remark 5.21. Notice that when $\sigma = 0$ the Dirac structures given in Eq.(5.54) and Eq.(5.53) are the same as the one given in Eq.(5.51) meaning that when $\sigma = 0$ the extended port-Hamiltonian system behaves as a fixed parameters port-Hamiltonian system.

Remark 5.22. Some extra dissipating element can be added to dissipate energy stored in the parametric part of the state. In fact if we connect to the parametric power ports a dissipative element, the energy absorption leads to a decrease of the parametric part of the state. This strategy can be used to decrease parametric states independently of the energy flowing along the extended Dirac structure.

5.5.2 Parameters Associated to Energy Dissipation and Interconnection

Parameters relative to energy dissipation play the role of modulating energy absorption. Changing values of these parameters is safe since different values would imply only a different rate of absorption. There is only one thing to be aware of: as reported in Sec. 1.4, the power flow towards dissipative elements must always be positive and, therefore, we cannot take a set of parameters which makes this power flow negative. If this happened dissipative elements would be transformed in power injecting elements which, therefore, would destroy passivity of the port-Hamiltonian system. This constraint in the choice of dissipative parameters is not very restrictive. In fact, in several applications (e.g. telemanipulation), port-Hamiltonian controllers are used to reproduce their physical equivalent and modulating physical parameter has the aim to obtain the same kind of behavior but characterized by different physical properties. Transforming, by a certain choice of parameters, dissipation in energy production we obtain a totally different kind of behavior which is always an undesired feature.

Parameters relative to the interconnection modulation can be freely changed since they modulate transformations which are lossless independently of their modulation constant.

5.5.3 Simulations

The aim of this section is to provide some simulations in order to validate the obtained results. Consider the system shown in Fig. 5.7, where the human operator interacts with a device, modeled as a mass m, controlled by an port-Hamiltonian impedance controller equivalent to the mechanical parallel of a linear spring, characterized by a stiffness k, and of a damper, with damping coefficient b. In this simplified scheme, we assume that the value of the damping coefficient b is fixed, while the stiffness k can be adapted in order to change the sensation rendered to the human operator.

Suppose that we want to change the value of the stiffness of the impedance controller and to drive it to the value k_{ref}. The energy storage function of the controller is:

$$H(k,x) = \frac{1}{2}kx^2 \tag{5.56}$$

where the rest length of the spring is supposed to be zero. The power port expressing the elongation:

$$\left(\frac{\partial H}{\partial x}, -\dot{x}\right) = (kx, -\dot{x}) = (e_x, f_x) \tag{5.57}$$

The following *parametric port* describes the power flow related to the stiffness variation:

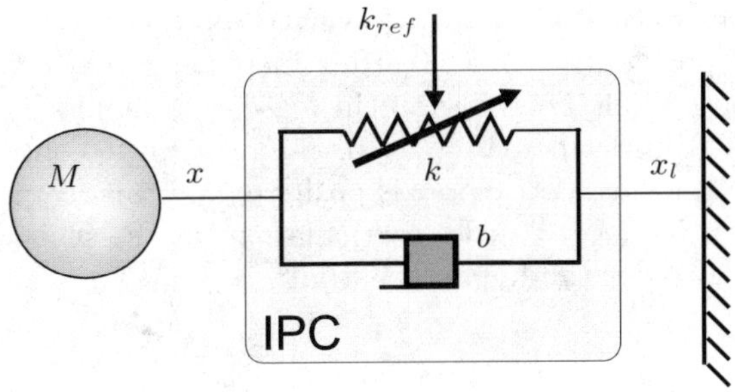

Fig. 5.7. A 1-DOF variable parameters port-Hamiltonian impedance controller

$$\left(\frac{\partial H}{\partial k}, -\dot{k}\right) = \left(\frac{x^2}{2}, -\dot{k}\right) = e_k, f_k) \tag{5.58}$$

Notice that the effort e_k is always positive and the sign of the power exchanged through the parametric port is determined only by the flow f_k. Finally, the impedance felt by the user is described by the *external* power port (e, f), and the goal of the stiffness adaptation is to compute an external effort e as close as possible to that generated by a real spring of stiffness k_{ref}:

$$f = \dot{x}, \qquad e = k_{ref}x \tag{5.59}$$

We choose to deviate part of the energy that is flowing towards the elongation port to the parametric port in order to passively change the stiffness of the spring; therefore, we extend the Dirac structure of the fixed parameters port-Hamiltonian system in the way reported in Eq.(5.53). This leads to:

$$f = \dot{x} \text{ and } \dot{k} = \sigma(t)f \tag{5.60}$$

where $\sigma(t)$ is the modulation law that has to be computed in order to achieve the desired behavior on the external port. Notice that the stiffness variation $\dot{k}$ is proportional to the external flow f and it takes place only when $\sigma(t) \neq 0$. This choice satisfies the first equation of Eq.(5.59) and straightforward computations show that the elongation perceived externally is equal to x. By expressing the Dirac structure in kernel form, we obtain the following equations:

$$\begin{pmatrix} 1 & 0 & 1 \\ 0 & 1 & \sigma \\ 0 & 0 & 0 \end{pmatrix} \begin{pmatrix} -\dot{x} \\ -\dot{k} \\ f \end{pmatrix} + \begin{pmatrix} 0 & 0 & 0 \\ 0 & 0 & 0 \\ 1 & \sigma & -1 \end{pmatrix} \begin{pmatrix} e_x \\ e_k \\ e \end{pmatrix} = 0 \tag{5.61}$$

First of all, consider the situation when the user is extracting power from the external port, namely $P = \langle e, f \rangle < 0$, while a stiffness increase is required $(k(t) < k_{ref})$. If $\sigma(t) \neq 0$, a fraction of P would be extracted from the parametric port; since e_k is always positive, power can be extracted from the parametric port only if $\dot{k} < 0$, thus increasing the deviation from the requested stiffness level. Therefore in this case $\sigma(t)$ should be set to zero. On the other side, a stiffness increase ($\dot{k} > 0$) can be achieved only if the power P_k flowing through the parametric port is positive, and this requires, the external power to be positive ($P > 0$). In particular, from the second and the third row of Eq.(5.61), under the hypothesis that:

$$\int_0^t f(\tau)d\tau = x(t) - x(0)$$

where $x(0)$ is the initial elongation of the spring, we obtain:

$$\sigma(t) = 2\frac{k_{ref}(t) - k(t)}{x(t)} \tag{5.62}$$

Straightforward computations show that this modulation law preserves not only the elongations of the springs but also their rest positions. Therefore, the equilibrium position of the system of Fig. 5.8 is not altered by the stiffness adaptation. However the choice of having $\dot{k}$ proportional to the external flow f causes the stiffness adaptation to take place only in *dynamic* conditions. In other words, when the user slowly moves the master device and then stops, it may happen that f is not sufficient to achieve the desired stiffness k_{ref}. In addition, there are no degrees of freedom to increase the convergence ratio of k to k_{ref}.

In order to test the proposed algorithm, simulations have been performed on the scheme of Fig. 5.7 where the communication channel has been disabled; in other words, the system behaves as a mass interconnected to a fixed frame by a parallel of a spring and a damper. In Fig. 5.8 it is shown how the stiffness set-point is tracked and how this affects the behavior of the master device, whose initial momentum is non null. The stiffness set-point k_{ref} periodically switches from a low value (10 N/m) to an higher one (100 N/m) and after an initial transient due to the fact that the energy available is not enough, $k(t)$ tracks the reference value. Moreover, this affects the oscillation frequency of the mass, that is higher for higher stiffness values (see Fig. 5.8(b)) .

5.6 A Scheme for Transparent Port-Hamiltonian Based Telemanipulation

In this section we will use the concepts developed so far to design a transparency optimized intrinsically passive telemanipulation scheme for port-Hamiltonian systems. The scheme is illustrated in Fig. 5.9

The scheme is represented in bond-graph notation. The half arrows, $\mathbb{C}$ elements and $\mathbb{R}$ elements represent energy exchange, energy storage and energy dissipation respectively. The arrows barred elements are variable elements. In

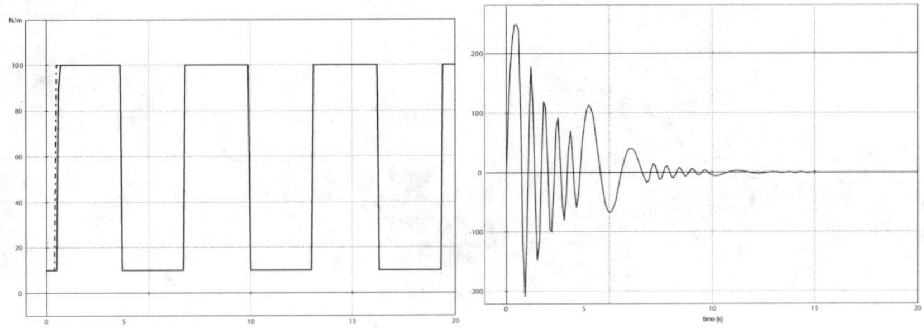

(a) k_{ref} (dotted line) and $k(t)$

(b) spring elongation

Fig. 5.8. Stiffness tracking and spring elongation

the scheme we have that master and slave robots are represented as port-Hamiltonian systems with fixed dissipations and fixed Dirac structures (depending on the physical structure of the systems) $\mathcal{D}_m$ and $\mathcal{D}_s$ respectively. The robots are controlled by means of variable parameters IPCs that are implemented by extended port-Hamiltonian systems. The communication between master and slave side is implemented by means of scattering variables and the characteristic impedance of the transmission line is Z. When the slave is interacting with the environment (which is assumed to be representable by the Hunt-Crossley model), the estimator derives the parameters characterizing the environment either by direct knowledge of using an identification algorithm. The estimated parameters are then transmitted to two signal processing subsystems which play the role of supervisors of the parametric part of state of the controllers. Given the parameters characterizing the remote environment, an optimal (i.e. maximizing transparency) set of parametric values is calculated for the IPCs. Then, in order to change the parameters, the supervisors enable some energy to flow towards the parametric ports and change the dissipation coefficients. The parametric variation is done preserving passivity as described in Sec. 5.5 and is stopped when the IPCs parameters have reached the desired set.

Remark 5.23. Since the estimator is at the slave side, there is some delay in the communication of the estimated parameters at the master side. Nevertheless, since we are transmitting *signals* and not power variables, no energy is produced and passivity is preserved. The only effect of delay is that the parametric modulation of the master IPC starts after the one of the slave IPC.

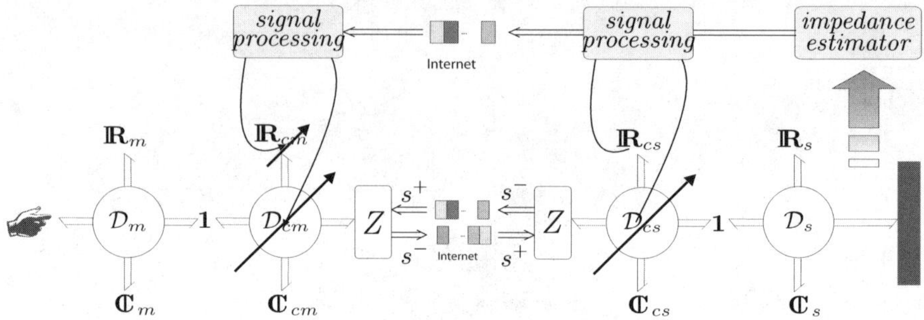

Fig. 5.9. Transparency optimized intrinsically passive telemanipulation scheme

5.7 Conclusions

It has been shown in Chap. 4 that the passivity theory and the port-Hamiltonian framework are very suitable for implementing bilateral telemanipulation systems which are characterized by a stable behavior both in case of free motion and in case of contact with any passive, possibly unknown, remote environment.

A stable behavior is just a first step for the implementation of a useful telerobotic system; the controllers and the communication channel should be designed in order to make the overall system as transparent as possible, namely to transmit to the user the feeling of directly interacting with the remote environment. In order to evaluate the kind of behavior that has to be reproduced at the master side, we have described the contact dynamics using the nonlinear Hunt-Crossley model.

One of the advantages of the port-Hamiltonian formalism is that it allows to consider general, possibly nonlinear, telemanipulators and, therefore, in order to preserve this degree of generality, we have exploited the behavioral approach to develop a framework for evaluating transparency, namely the match between the feeling perceived by the user and the real interaction taking place between the slave and the environment. The behavioral approach coupled with the port-Hamiltonian formalism, has provided a guideline for the analysis and the transparency based design of complex, possibly non linear, bilateral telemanipulation systems. The proposed framework allowed us to formally analyze the effect on transparency of the delay, of the packets loss and of the packets interpolation in scattering based packet switching communication channels.

Exploiting the port-Hamiltonian framework, we have designed nonlinear impedance controllers that mimic the behavior of the Hunt-Crossley model of the estimated environment, improving in this way the feeling of the remote interaction perceived by the user.

Since the slave can interact with several remote environments, it is necessary to be able to change the parameters characterizing the nonlinear impedance controllers. Exploiting the interconnection structure of the port-Hamiltonian systems, we have proposed a passivity preserving algorithm for tuning the controllers. This algorithm highlights the importance of the interconnection structure of port-Hamiltonian systems for the design of controllers. Some recent applications of port-Hamiltonian based telemanipulation where the Dirac structure of the controllers has been modified for improving performances can be found in [268, 269].

A

Mathematical Background

This appendix gives the necessary background on the mathematical objects used in the book. In particular, the concepts of manifold, tangent and cotangent bundles, tensors and Lie groups are defined and some of their properties are reported.

A.1 Manifolds and Vector Bundles

In this section the concepts of manifolds, tangent and cotangent bundle and vector bundle are introduced. For a more complete treatment the reader is addressed to [185, 73].
Roughly speaking, manifolds are abstract surfaces that locally look like (i.e. are locally diffeomorphic with) $\mathbb{R}^n$.

Definition A.1 (Coordinate Chart). *Given a set M, a* chart *on M is a subset U of M together with a bijective map $\phi : U \to \phi(U) \subset \mathbb{R}^n$. Usually $\phi(m)$ is denoted by $(x^1, \ldots, x^n)$ and x^i are called the* coordinates *of the point $m \in U \subset M$. A chart will be indicated as (U, ϕ).*

A coordinate chart allows to identify a subset of M with $\mathbb{R}^n$. Consider two coordinate charts (U, ϕ) and (U', ϕ') where $U \cap U' \neq \emptyset$. All the points of $U \cap U'$ are described by two coordinate charts, namely the two coordinate charts overlap.

Definition A.2 (Compatible Charts). *Two charts (U, ϕ) and (U', ϕ') such that $U \cap U' \neq \emptyset$ are called* compatible *if $\phi(U \cap U')$ and $\phi\prime(U \cap U')$ are open subsets of $\mathbb{R}^n$ and the maps:*

$$\phi' \circ \phi^{-1}\big|_{\phi(U \cap U')} : \phi(U \cap U') \to \phi'(U \cap U') \tag{A.1}$$

and

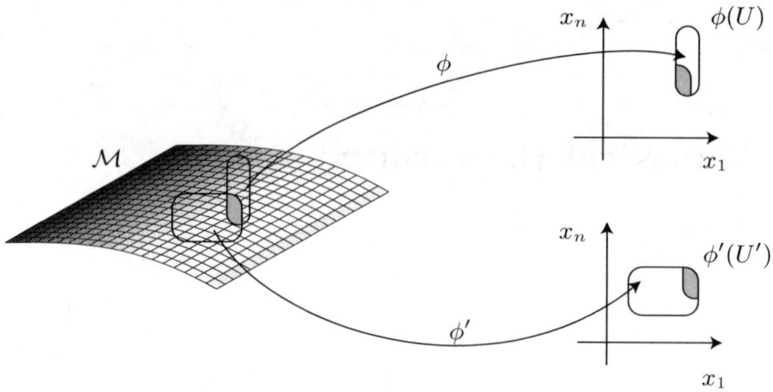

Fig. A.1. Compatible charts

$$\phi \circ (\phi')^{-1}|_{\phi'(U \cap U')} : \phi'(U \cap U') \to \phi(U \cap U') \tag{A.2}$$

are smooth.

In Fig. A.1 the concept of compatible charts is illustrated. It is now possible to define a particular collection of charts:

Definition A.3 (Atlas). *An* atlas *on a set* M *is a family of charts* $\mathcal{A} = \{(U_i, \phi_i) \mid i \in I\}$, *where* I *is an index set, such that:*

- $M = \cup_{i \in I} U_i$
- *For any* $i, j \in I$ *the charts* (U_i, ϕ_i) *and* (U_j, ϕ_j) *are compatible.*

Definition A.4 (Differentiable manifold). *Two atlases* $\mathcal{A}_1$ *and* $\mathcal{A}_2$ *are* equivalent *if* $\mathcal{A}_1 \cup \mathcal{A}_2$ *is an atlas. A* differentiable structure D *on* M *is an equivalence class of atlases on* M. *The union of all the atlases in* D,

$$\mathcal{A}_D = \bigcup \{\mathcal{A} \,|\, \mathcal{A} \in D\}$$

is the maximal atlas *of* D *and a chart* $(U, \phi) \in \mathcal{A}_D$ *is an* admissible chart. *A* differentiable manifold $\mathcal{M}$ *is a pair* (M, D), *where* M *is a set and* D *is a differentiable structure on* M.

The set M defines the set of points of the manifold, while the atlas, namely the way used to map the points to $\mathbb{R}^n$, defines the differential structure and allows differential calculus over the manifold. Atlases belonging to different equivalence classes define different differentiable structures which yield to completely different manifolds.

Definition A.5 (Dimension of a differentiable manifold). *The* dimension *of a manifold* $\mathcal{M}$ *around a point* $p \in U_i \subset M$ *is the dimension of the linear space which is the co-domain of the map* ϕ_i *of the chart* (U_i, ϕ_i). *If the dimension around each point of the manifold is the same, for instance* n, *then* n *is said to be the dimension of the manifold.*

Consider two curves c_1, c_2 that takes value over a manifold $\mathcal{M}$, namely:

$$c_1, c_2 : \mathbb{R} \to \mathcal{M}$$

The curves are said to be *equivalent* if

$$c_1(0) = c_2(0) = m \quad \text{and} \quad (\phi \circ c_1)'(0) = (\phi \circ c_2)'(0) \tag{A.3}$$

in some chart ϕ; the prime indicates the usual derivative in $\mathbb{R}^n$. It is possible to prove that this definition of equivalence is chart independent and that it defines an equivalence relation. All the curves in the same equivalence class have the same tangent, calculated using the differential structure of $\mathcal{M}$, at the point m.

Definition A.6 (Tangent vector). *A* tangent vector v *to a manifold* $\mathcal{M}$ *in a point* $m \in \mathcal{M}$ *is an equivalence class of curves at* m. *The components of* v *are the numbers* $v^1, \ldots, v^n$ *defined by:*

$$v^i = \frac{d}{dt}(\phi \circ c)^i\big|_{t=0} \tag{A.4}$$

where $i = 1 \ldots n$ *and* c *is a representative curve of the equivalence class defining* v *and* ϕ *is a chart of* $\mathcal{M}$.

It is possible to prove that the set of tangent vectors at a point $m \in \mathcal{M}$ forms a vector space.

Definition A.7 (Tangent space). *The vector space formed by all the tangent vectors at a point* $m \in \mathcal{M}$ *is called* tangent space *to* $\mathcal{M}$ *at* m *and is indicated by* $T_m\mathcal{M}$.

Definition A.8 (Tangent bundle). *The* tangent bundle *of* $\mathcal{M}$, *denoted by* $T\mathcal{M}$, *is the set given by the union of all the tangent spaces to* $\mathcal{M}$ *at the points* $m \in \mathcal{M}$, *that is:*

$$T\mathcal{M} = \bigcup_{m \in \mathcal{M}} T_m\mathcal{M} \tag{A.5}$$

Thus a point of $T\mathcal{M}$ *is a vector that is tangent to* $\mathcal{M}$ *in some point* $m \in \mathcal{M}$.

It is possible to define the dual concepts of tangent space and tangent bundle:

Definition A.9 (Cotangent space). *The vector space formed by the set of linear operators from* $T_m\mathcal{M}$ *to* $\mathbb{R}$, *where* $m \in \mathcal{M}$, *is called* cotangent space *to* $\mathcal{M}$ *at* m *and is indicated by* $T^*_m\mathcal{M}$.

An element $v^* \in T^*_m\mathcal{M}$ is called cotangent vector or simply *covector*.

Definition A.10 (Cotangent bundle). *The* cotangent bundle *of $\mathcal{M}$, denoted by $T^*\mathcal{M}$, is the set given by the union of all the cotangent spaces to $\mathcal{M}$ at the points $m \in \mathcal{M}$, that is:*

$$T^*\mathcal{M} = \bigcup_{m \in \mathcal{M}} T^*_m\mathcal{M} \tag{A.6}$$

Thus a point of $T\mathcal{M}$ is a linear operator on a tangent space of $\mathcal{M}$ at some point $m \in \mathcal{M}$.

Once the concepts of tangent space and cotangent space have been defined, it is possible to define operators that associate to the points of a manifold, element in the tangent or in the cotangent space.

Definition A.11 (Vector field). *A* vector field *X on a manifold $\mathcal{M}$ is a map $X : \mathcal{M} \to T\mathcal{M}$ that assigns a tangent vector $X(m) \in T_m\mathcal{M}$ at the point $m \in \mathcal{M}$. An* integral curve *with initial condition m_0 at $t = 0$ is a differentiable map $c : (a, b) \to \mathcal{M}$ such that (a, b) is an open interval containing 0, $c(0) = m_0$ and*

$$c'(t) = X(c(t)) \quad \forall t \in (a, b) \tag{A.7}$$

Thus, it is possible to interpret a vector field as an operator that associates to each point of the manifold a "velocity vector". The integral curve of a vector field is a curve whose velocity at each point is that associated by the vector field to that point. The dual concept of a vector field is that of covector field and it is defined as:

Definition A.12 (Covector field). *A* covector field *X^* on a manifold $\mathcal{M}$ is a map $X^* : \mathcal{M} \to T^*\mathcal{M}$ that assigns a covector $X^*(m) \in T^*_m\mathcal{M}$ at the point $m \in \mathcal{M}$.*

Consider a smooth map f defined on a manifold $\mathcal{M}$, i.e.:

$$f : \mathcal{M} \to \mathbb{R}$$

and let X be a vector field defined over $\mathcal{M}$. It is possible to consider the variation of the map f along the direction determined by X, i.e. along the integral curves of X.

Definition A.13 (Lie derivative). *The derivative of the function f along the vector field X is a map defined on $\mathcal{M}$ and it is defined by:*

$$L_X f : \mathcal{M} \to \mathbb{R} \qquad L_X f(x) = \frac{\partial^T f}{\partial x} X(x) \quad x \in \mathcal{M} \tag{A.8}$$

where

$$\frac{\partial f}{\partial x} = [\frac{\partial f}{\partial x_1}, \ldots, \frac{\partial f}{\partial x_n}]^T$$

Given two vector fields X and Y on a manifold $\mathcal{M}$, it is possible to define a third vector field through the Jacobi-Lie brackets.

Definition A.14 (Jacobi-Lie brackets). *Given two vector field X and Y on a manifold $\mathcal{M}$ and a map $f : \mathbb{R} \to \mathcal{M}$, it is possible to define a third vector field, which is called the* Jacobi-Lie brackets *of X and Y and which is indicated with $[X, Y]$ and which is unique, such that:*

$$L_{[X,Y]}f = L_X L_Y f - L_Y L_X f \tag{A.9}$$

The Jacobi-Lie brackets of two vector fields depends on the commutativity of the starting vector fields. In fact it is not said that moving forward along an integral curve of X and then moving forward along an integral curve of Y and then moving backward along an integral curve of X and finally moving backward along an integral curve of Y the starting position is reached again. The Jacobi-Lie brackets of X and Y express this non commutativity.

Definition A.15 (Tangent map). *Given two manifolds $\mathcal{M}$ and $\mathcal{N}$ and a smooth map $f : \mathcal{M} \to \mathcal{N}$, it is possible to define a linear map:*

$$T_m f : T_m\mathcal{M} \to T_{f(m)}\mathcal{N} \tag{A.10}$$

constructed in the following way. For $v \in T_m\mathcal{M}$, choose a curve $c : (-\epsilon, \epsilon) \to \mathcal{M}$ with $c(0) = m$ and velocity vector $dc/dt|_{t=0} = v$. Then $T_m f(v)$ is the velocity vector at $t = 0$ of the curve $f \circ c : \mathbb{R} \to \mathcal{N}$, that is,

$$T_m f(v) = \frac{d}{dt} f(c(t))\big|_{t=0} \tag{A.11}$$

The vector $T_m f(v)$ does not depend on the curve c but only on the vector v. The tangent map *of f is the map:*

$$Tf : T\mathcal{M} \to T\mathcal{N} \quad (m, v) \to (f(m), T_m f(v)) \tag{A.12}$$

A generalization of the concepts of vector and covector fields are the concept of distribution and codistribution.

Definition A.16 (Distribution). *A distribution Δ on a n-dimensional differentiable manifold $\mathcal{M}$ is a smooth map that assigns to each point $m \in \mathcal{M}$ a subspace $\Delta(m) \subset T_m\mathcal{M}$. A distribution is called* smooth *if and only if the subspaces that it determines on the tangent space of each point can be spanned by a set of smooth vector fields, i.e. there exist $p(\leq n)$ smooth vector fields X_i such that:*

$$\Delta(m) = span\,\{X_i(m)\} \quad i = 1, \dots, p$$

The distribution is called constant dimensional if and only if, for each $m \in \mathcal{M}$, $\Delta(m)$ has the same dimension.

Definition A.17 (Codistribution). *A codistribution Δ^* on a n-dimensional differentiable manifold $\mathcal{M}$ is a smooth map that assigns to each point $m \in \mathcal{M}$ a subspace $\Delta^*(m) \subset T^*_m\mathcal{M}$. A codistribution is called* smooth *if and only if the subspaces that it determines on the cotangent space of each point can be spanned by a set of smooth covector fields, i.e. there exist $p(\leq n)$ smooth vector fields X^*_i such that:*

$$\Delta^*(m) = span\,\{X^*_i(m)\} \quad i = 1, \ldots, p$$

The codistribution is called constant dimensional if and only if, for each $m \in \mathcal{M}$, $\Delta(m)$ has the same dimension.

Another concept relative to differentiable manifolds that is used in the book is that of (smooth) vector bundle. Roughly speaking, a vector bundle is a manifold with a vector space attached to each point. Tangent and cotangent bundles to a manifold are examples of vector bundles. More formally, let $\mathcal{M}$ be a manifold:

Definition A.18 (Vector bundle). *A* vector bundle *ξ over $\mathcal{M}$ consists of:*

1. *A manifold $\mathcal{E}$, called the* total space
2. *A smooth map $\pi : \mathcal{E} \to \mathcal{M}$, called* projection map
3. *For each $m \in \mathcal{M}$, the set $\pi^{-1}(m)$ has the structure of a vector space.*

Furthermore, the following condition of local triviality, *must be satisfied: for each $m \in \mathcal{M}$ there exists a neighborhood $U \in \mathcal{M}$, an integer $n > 0$ and a diffeomorphism*

$$h : U \times \mathbb{R}^n \to \pi^{-1}(U)$$

such that, for each $m \in U$, the correspondence $x \to h(m, x)$ defines an isomorphism between the vector space $\mathbb{R}^n$ and the vector space $\pi^{-1}(m)$. The pair (U, h) is called a local coordinate system for ξ. The vector space $\pi^{-1}(m)$ is called fiber *over m for any $m \in \mathcal{M}$.*

For a very complete treatment of vector bundles see [204].

A.2 Tensors

Tensors are the generalization of the concepts of vector and matrix and, in fact, once a coordinate frame has been defined, they can be represented as multidimensional matrices. For a complete treatment see, for example, [185].

Definition A.19 (Multilinear map). *Consider $n+1$ vector spaces $\mathcal{V}_1, \ldots, \mathcal{V}_n, \mathcal{W}$. A* multilinear map *is a map:*

$$L : \mathcal{V}_1 \times \cdots \times \mathcal{V}_n \to \mathcal{W} \tag{A.13}$$

such that, for each $i = 1, \dots, n$, the maps:

$$L_i(v_1, \dots, v_{i-1}, v_{i+1}, \dots, v_n) : \mathcal{V}_i \to \mathcal{W} \quad v_i \to L(v_1, \dots, v_n) \tag{A.14}$$

are linear for each $v_1, \dots, v_{i-1}, v_{i+1}, \dots, v_n$.

The set of multilinear operators defined on $\mathcal{V}_i, \dots, \mathcal{V}_n, \mathcal{W}$ is indicated as

$$L^n(\mathcal{V}_1, \dots, \mathcal{V}_n; \mathcal{W})$$

and it can be proven that it is a vector space.
Consider a vector space $\mathcal{V}$ and denote its dual space, namely the vector space formed by the linear operators on $\mathcal{V}$, by $\mathcal{V}^*$.

Definition A.20 (Tensors). *Given a vector space $\mathcal{V}$, the set of tensors of order $p+q$ of type (p, q), namely p contravariant and q covariant, is defined by:*

$$\mathcal{T}_q^p(\mathcal{V}) = L^{p+q}(\underbrace{\mathcal{V}, \dots, \mathcal{V}}_{q\,times}, \underbrace{\mathcal{V}^*, \dots \mathcal{V}^*}_{p\,times}; \mathbb{R}) \tag{A.15}$$

Notice that $\mathcal{T}_0^1(\mathcal{V}) = \mathcal{V}$ and that $\mathcal{T}_1^0(\mathcal{V}) = \mathcal{V}^*$.
It is possible to define tensors over a differentiable manifold. In Sec. A.1 it has been shown that it is possible to define a vector space (the tangent space) at each point of a manifold; this vector space can be used as the linear space over which to define a tensor. More formally, it is possible to give the following:

Definition A.21 (Tensor field). *A* p contravariant q covariant tensor field *is a map that smoothly assigns to each point m of a manifold $\mathcal{M}$ a tensor in $\mathcal{T}_q^p(T_m\mathcal{M})$*

It is then possible to associate to each point m of a manifold $\mathcal{M}$ the vector space of (p, q) tensors on $T_m\mathcal{M}$; this vector space is indicated with $T_{q,p}\mathcal{M}_m$.

Definition A.22 (Tensor bundle). *The q contravariant and p covariant tensor bundle over a manifold $\mathcal{M}$ is defined as:*

$$\mathcal{T}_{q,p}\mathcal{M} = \bigcup_{m \in \mathcal{M}} T_{q,p}\mathcal{M}_m \tag{A.16}$$

The concepts of vector and covector fields and of tangent and cotangent bundles are special cases of the concepts of tensor field and tensor bundle.

A.3 Lie Groups and Rigid Motions

The concept of Lie group is very useful in robotics and, in general, in mechanics, because it allows to describe the motion of rigid bodies and mechanisms in a very elegant and coordinate free way. For an excellent and complete treatment of Lie groups and their applications see [221].

Definition A.23 (Lie group). *A* Lie group *is a manifold $\mathcal{G}$ that is also a group. The group structure must be compatible with the differential structure of manifold, that is, the group multiplication*

$$\mu : \mathcal{G} \times \mathcal{G} \rightarrow \mathcal{G} \quad (g, h) \rightarrow gh$$

is a smooth map.

Since the manifold $\mathcal{G}$ is a group it is possible to identify the identity element which is indicated with e. Furthermore, since the group $\mathcal{G}$ is a manifold, it is possible to define tangent and cotangent spaces at each point $g \in \mathcal{G}$.

There are two important mappings of Lie groups:

Definition A.24 (Left and right translations). *Given a Lie group $\mathcal{G}$, for each $g \in \mathcal{G}$ the* left translation by g *is defined by:*

$$L_g : \mathcal{G} \rightarrow \mathcal{G} \quad h \rightarrow gh \tag{A.17}$$

and the right translation by g *is defined by:*

$$R_g : \mathcal{G} \rightarrow \mathcal{G} \quad h \rightarrow hg \tag{A.18}$$

Definition A.25 (Lie algebra). *A Lie algebra $\mathfrak{g}$ is a vector space together with a bilinear antisymmetric bracket satisfying Jacobi's identity, that is, for each $\xi, \eta . \zeta \in \mathfrak{g}$*

$$[[\xi, \eta], \zeta] + [[\zeta, \xi], \eta] + [[\eta, \zeta], \xi] = 0 \tag{A.19}$$

For each $\xi \in T_e\mathcal{G}$, it is possible to define a vector field X_ξ on $\mathcal{G}$ by letting:

$$X_\xi(g) = T_e L_g(\xi) \tag{A.20}$$

Definition A.26 (Lie algebra associated to a Lie group). *Each Lie group $\mathcal{G}$ is associated to an unique Lie algebra $\mathfrak{g}$. The vector space relative to $\mathfrak{g}$ is $T_e\mathcal{G}$ and the commutator of the algebra is given by the Lie brackets, defined as follows: $\forall \xi, \eta \in T_e\mathcal{G}$*

$$[\xi, \eta] = [X_\xi, X_\eta]$$

where $[X_\xi, X_\eta]$ is the Jacobi-Lie brackets (Def. A.14) between vector fields on a manifold.

A.3.1 An Example: The Special Euclidean Group SE(3)

The special Euclidean group $SE(3)$ is an example of Lie group that is also very useful in the study of the motion of rigid bodies and mechanisms.

It is well known that the configuration of a rigid body in the space can be described, using homogeneous coordinates, by the homogeneous matrix:

$$h = \begin{pmatrix} R & p \\ 000 & 1 \end{pmatrix} \tag{A.21}$$

where R is a 3×3 orthogonal matrix with determinant equal to 1 which represents the orientation of the rigid body while p is a vector that represent its position. The set of matrices of this form is called special Euclidean group in $\mathbb{R}^3$ and is denoted by $SE(3)$. It can be easily seen that the set $SE(3)$ has the structure of manifold. The structure of group can be checked taking as operation the matrix multiplication. Thus, $SE(3)$ is both a manifold and a group and, therefore, it is a Lie group.

Consider two rigid bodies i and j, the relative configuration of i and j is represented by an homogeneous matrix h_i^j. The Lie algebra of $SE(3)$ is denoted by $\mathfrak{se}(3)$ and it is the set of matrices of the form

$$\begin{pmatrix} \omega & v \\ 000 & 0 \end{pmatrix} \tag{A.22}$$

where ω is 3×3 skew-symmetric matrix and $v \in \mathbb{R}^3$. The Lie brackets of the Lie algebra are given by:

$$[A, B] = AB - BA \quad \forall A, B \in \mathfrak{se}(3) \tag{A.23}$$

An element of $\mathfrak{se}(3)$ is called a *twist*. The vector space $\mathfrak{se}(3)$ has dimension 6 and is isomorphic to $\mathbb{R}^6$.
The relative motion of two rigid bodies can be described by a curve $h_i^j(t)$ on $SE(3)$. The generalized relative velocity is $\dot{h}_i^j \in T_{h_i^j(t)} SE(3)$.

The most general instantaneous motion of a rigid body in the space is a screw motion (Chasles' theorem, [290, 272]), i.e. a rototranslation around an instantaneous axis in the space. Thanks to the tangent maps of the left (or right) translation map it is possible to describe the generalized velocity through a twist. In particular $t_i^j \in \mathfrak{se}(3)$ represents the instantaneous velocity of the motion, the skew symmetric matrix ω represent the rotation around the instantaneous axis and v the velocity along the instantaneous axis.
Dually, the most general instantaneous system of forces that can be applied to a rigid body is given by a pure momentum and a pure translation around an instantaneous spatial axis (Poinsot's theorem, [290, 272]). Using duality concepts, it is possible to express such an instantaneous system of forces by an element of $\mathfrak{se}^*(3)$, the vector space given by the linear operators acting on $\mathfrak{se}(3)$. Elements of $\mathfrak{se}^*(3)$ are called *wrenches* and w_i represents the wrench applied to the body i. For further details see [290, 272, 212].

References

1. J.J. Abbot and A.M. Okamura. Analysis of virtual fixture contact stability for telemanipulation. In *Proceedings of IEEE/RSJ International Conference on Intelligent Robots and Systems*, Las Vegas, NV, USA, 2003.
2. J.A. Acosta, R. Ortega, A. Astolfi, and A.D. Mahindrakar. Interconnection and damping assignment passivity-based control of mechanical systems with underactuation degree one. *IEEE Transactions on Automatic Control*, 50(12):1936–1955, December 2005.
3. R.J. Adams and B. Hannaford. Stable haptic interaction with virtual environnment. *IEEE Transactions on robotics and Automation*, 1999.
4. R.J. Adams and B. Hannaford. Control law design for haptic interfaces to virtual reality. *IEEE Transactions on Control Systems Technology*, 10(1), January 2002.
5. R.J. Adams, M.R. Moreyra, and B. Hannaford. Stability and performance of haptic displays: Theory and expreiments. In *Proceedings ASME International Mechanical Engineering Congress and Exhibition*, Anaheim, California, 1998.
6. N. Aghannan and P. Rouchon. An intrinsic observer for a class of Lagrangian systems. *IEEE Transactions on Automatic Control*, 48(6), June 2003.
7. R. Anderson and M. Spong. Bilateral control of teleoperators with time delay. *IEEE Transactions on Automatic Control*, 34(5):494–501, 1989.
8. R.J. Anderson and M.W. Spong. Asymptotic stability for force reflecting teleoperators with time delays. In *Proceedings of IEEE International Conference on Robotics and Automation*, May 1989.
9. P. Arcara and C. Melchiorri. A comparison of control schemes for teleoperation with time delay. In *Proceedings of IFAC TA 2001, Conf. on Telematics Applications*, 2001.
10. P. Arcara and C. Melchiorri. Control schemes for teleoperation with time delay: a comparative study. *Robotics and Autonomous Systems*, 38(1), 2002.
11. P. Arcara and C. Melchiorri. Position drift compensation for a passivity-based telemanipulation control scheme. In *Proceedings to Mechatronics Conference*, Enschede, The Netherlands, June 2002.
12. S. Arimoto and F. Miyazaki. Stability and robustness of pid feedback control for robot manipulators of sensory capability. In M. Brady and R. Paul, editors, *Robotics Res.: The First Int. Symp.*, pages 783–799. MIT Press, 1984.

13. H. Arioui, A. Kheddar, and S. Mammar. A predictive wave-based approach for time delayed virtual environments haptics systems. In *Proceedings of the* 2002 *IEEE Int. Workshop on robot and Human Interactive Communication*, Berlin, Germany, September 2002.
14. V.I. Arnold. *Mathematical Methods in Classical Mechanics.* Springer-Verlag, New York, 2nd edition edition, 1989.
15. A. Astolfi, A. Isidori, and L. Marconi. A note on disturbance suppression for hamiltonian systems by state feedback. In *Proceedings of IFAC Workshop on Lagrangian and Hamiltonian Methods for Nonlinear Control*, Sevilla, Spain, April 2003.
16. T. Bastogne. Behavioral descriptions of the object-oriented paradigm for physical system modeling. In *Proceedings of IFAC world Congress*, Prague, Czech Republic, July 2005.
17. C. Batlle, A. Doria-Cerezo, and E. Fossas. Ida-pbc controller for a bidirectional power flow full-bridge rectifier. In *Proceedings of IEEE Conference on Decision and Control*, Seville, Spain, December 2005.
18. C. Batlle, Arnau Doria-Cerezo, and R. Ortega. Power flow control of a doubly fed induction machine coupled to a flywheel. In *Proceedings of IEEE Conference on Control Applications*, Taipei, Taiwan, 2004.
19. C. Benedetti, M. Franchini, and P. Fiorini. Stable tracking in variable time-delay teleoperation. In *Proceedings of IEEE/RSJ International Conference on Intelligent Robots and Systems*, Maui, Hawaii, USA, October 2001.
20. P. Berestesky, N. Chopra, and M.W. Spong. Discrete time passivity in bilateral teleoperation over the internet. In *Proceedings of IEEE International Conference on Robotics and Automation*, New Orleans, Louisiana, 2004.
21. A. Bettini, P. Marayong, S. Lang, A.M. Okamura, and G.D. HAger. Vision-assisted control for manipulation using virtual fixtures. *IEEE Transactions on Robotics*, 20:953–966, 2004.
22. A. Bicchi, C. Melchiorri, and D. Ballucchi. On the mobility and manipulability of multiple whole-limb robots. *IEEE Transactions on Robotics and Automation*, 11(2), 1995.
23. A. Bicchi, E.P. Scilingo, and D. De Rossi. Haptic discrimination of softness in teleoperation: The role of the contact area spread rate. *IEEE Transactions on Robotics and Automation*, 16(5), 2000.
24. G. Blankenstein and A.J. van der Schaft. Symmetry and reduction in implicit generalized Hamiltonian systems. *Reports on mathematical physics*, 47(1), 2001.
25. A.M. Bloch and P.E. Crouch. Representation of dirac structures on vector spaces and nonlinear lcv-circuits. In H.G. Ferraya, R. Gardner, and H. Sussman Eds., editors, *Proc. of Symposia in Pure Mathematics and, Differential Geometry and Control Theory*, volume 64, pages 103–117, 1999.
26. M. Bonfé, C. Fantuzzi, and C. Secchi. Inheritance of behavior in object-oriented designs for industrial control systems. In *Proceedings of IFAC world Congress*, Prague, Czech Republic, July 2005.
27. M. Bonfé, C. Fantuzzi, and C. Secchi. Object-oriented modeling of multi-domain systems. In *Proceedings of IEEE Conference on Automation Science and Engineering*, Edmonton, Canada, August 2005.
28. M. Bonfé, C. Fantuzzi, and C. Secchi. Uml come linguaggio di modellazione unificato per sistemi meccatronici. In *Covegno ANIPLA su Automazione per lo sviluppo e la competitività delle PMI*, Napoli, Italia, November 2005.

29. M. Bonfé, C. Fantuzzi, and C. Secchi. Unified modeling and verification of logic controllers for physical systems. In *Proceedings of IEEE Conference on Decision and Control*, Seville, Spain, December 2005.
30. M. Bonfé, C. Fantuzzi, and C. Secchi. Verification of behavioral substitutability in object-oriented models for industrial controllers. In *Proceedings of IEEE International Conference on Robotics and Automation*, Barcelona, Spain, April 2005.
31. C. Bonivento, L. Marconi, and R. Zanasi. Output regulation of nonlinear systems by sliding mode. *Automatica*, 2001.
32. W. Borutzky. Bond graph modeling from an object oriented modeling point of view. *Simulation practice and theory*, 7:439–461, 1999.
33. W. Borutzky, B. Barnard, and J.U. Thoma. Describing bond graphs models of hydraulic components in modelica. In *Proceedings of IMACS Symposium on Mathematica Modeling*, pages 715–719, Vienna, Austria, February 2000.
34. D. Botturi, A. Castellani, D. Moschini, and P. Fiorini. Performance evaluation of task control in teleoperation. In *Proceedings of IEEE International Conference on Robotics and Automation*, New Orleans, LA, USA, April 2004.
35. M. Boukhnifer, A. Ferreira, and J.-G. Fontaine. Scaled teleoperation controller design for micromanipulation over internet. In *Proceedings of IEEE International Conference on Robotics and Automation*, New Orleans, LA, USA, April 2004.
36. P.C. Breedveld. *Physical systems theory in terms of Bond-Graphs*. PhD thesis, University of Twente, 1984.
37. P.C. Breedveld. An alternative model for static and dynamic friction in dynamic system simulation. In *Proceedings to Mechatronics Forum International Conference*, volume 2, pages 717–722, Darmstadt, Germany, September 2000.
38. J.M. Brown and J.E. Colgate. Minimum mass for haptic display simulations. In *Proceedings of the ASME Interational Mechanical Engineering Congress and Exhibition*, 1998.
39. P. Buttolo and B. Hannaford. Advantages of actuation redundancy for the design of haptic displays. In *Proceedings ASME, Fourth Annual Symposium on Haptic Interfaces for Virtual Environment and Teleoperator Systems*, S. Francisco, November 1995.
40. P. Buttolo, D. Kung, and B. Hannaford. Manipulation in real, virtual and remote environments. In *Proceedings of IEEE International Conference on Systems, Man and Cybernetics*, Vancouver, Canada, October 1995.
41. C.I. Byrnes, A. Isidori, and J.C. Willems. Passivity, feedback equivalence and the global stabilization of minimum phase nonlinear systems. *IEEE Transactions on Automatic Control*, 36(11), 1991.
42. L. Capretz. A brief history of the object-oriented approach. *ACM SIGSOFT Software Engineering Notes*, 28(2), 2003.
43. A. Casavola, M. Sorbara, and S. Stramigioli. A passivity-based command governor control approach. In *Proceedings of the IFAC Symposium on Robot Control*, Bologna, Italy, September 2006.
44. J. Cervera and A.J. van der Schaft. *On composition of Dirac strctures and its implications for control by interconnection*, chapter Nonlinear and adaptive control. Springer Lecture notes in Control and Informations Sciences, 2002.
45. C.C. Cheah, M. Hirano, S. Kawamura, and S. Arimoto. Approximate jacobian control for robots with uncertain kinematics and dynamics. *IEEE Transactions on Robotics and Automation*, 19(4), August 2003.

46. Y. Chen and K.L. Moore. Analytical stability bound for delayed second-order systems with repeating poles using Lambert function W. *Automatica*, 38:891–895, 2002.
47. C-C Cheng and J-H Chen. A generalized control approach for ideal teleoperation. In *Proceedings of IEEE/ASME International Conference on Advanced Intelligent Mechatronics*, Como, Italy, July 2001.
48. C. Cho, C.-S. Hwang, J.-B. Song, and M. Kim. Stable haptic display of slowly updated virtual environment with multirate wave transform. In *Proceedings of IEEE International Conference on Robotics and Automation*, Barcelona, Spain, April 2005.
49. H.C. Cho and J.H. Park. Impedance controller design of internet-based teleoperation using absolute stability concept. In *Proceedings to IEEE/RSJ Conference on Intelligent Robots and Systems*, EPFL, Lausanne, Switzerland, October 2002.
50. S. Choi, D. Park, and S Jayasuria. A time-varying sliding surface for fast and robust tracking control of second order uncertain systems. *Automatica*, 30(5), 1994.
51. N. Chopra, M.W. Spong, S. Hirche, and M. Buss. Bilateral teleoperation over the internet: the time varying delay problem. In *Proceedings of American Control Conference*, volume 1, Denver, Colorado, USA, June 2003.
52. N. Chopra, M.W. Spong, and R. Lozano. Adaptive coordination control of bilateral teleoperators with time delay. In *Proceedings of IEEE Conference on Decision and Control*, Paradise Island, Bahamas, December 2004.
53. L. O. Chua. Memristor - the missing circuit element. *IEEE Transactions on Circuit Theory*, CT-18(5), September 1971.
54. H. Chyi and Y.-C. Cheng. A note on the use of the Lambert W function in the stability analysis of time-delay systems. *Automatica*, 41:1979–1985, 2005.
55. C.L. Clover. A control-system architecture for robots used to simulate dynamic force and moment interaction between humans and virtual objects. *IEEE Transactions on Systems, Man and Cybernetics*, 1999.
56. J.E. Colgate. Power and impedance scaling in bilateral manipulation. In *Proceedings of IEEE International Conference on Robotics and Automation*, Sacramento, California, April 1991.
57. J.E. Colgate. Robust impedance shaping telemanipulation. *IEEE Transactions on Robotics and Automation*, 1993.
58. J.E. Colgate and J.M. Brown. Factors affecting the z-width of a haptic display. In *Proceedings to IEEE International Conference on Robotics and Automation*, San Diego, California, May 1994.
59. J.E. Colgate, P.E. Grafing, M. C. Stanley, and G. Schenkel. Implementation of stiff virtual walls in force reflecting interfaces. In *Virtual Reality Annual International Symposium*, 1993.
60. J.E. Colgate and G. Schenkel. Passivity of a class of sampled-data systems: Application to haptic interfaces. In *Proceedings to American Control Conference*, 1994.
61. J.E. Colgate, M.C. Stanley, and J.M. Brown. Issues in the haptic display of tool use. In *Proceedings of IEEE/RSJ International Conference on Intelligent Robots and Systems*, 1995.
62. N. Copra, M.W. Spong, R. Ortega, and N.E. Barabanov. On position tracking il bilateral teleoperation. In *Proceedings of American Control Conference*, Boston, Massachussets, June 2004.

63. J. Cortés, A.J. van der Schaft, and P.E. Crouch. Gradient realization of nonlinear control systems. In *Proceedings of IFAC Workshop on Lagrangian and Hamiltonian Methods for Nonlinear Control*, Seville, Spain, April 2003.
64. F. Couenne, D. Eberard, C. Jallut, L. Lefevre, and B. Maschke. Multiscale distributed parameter model of an adsorption column using a port-based geometric approach. In *Proceedings of the European Symposium on Computer Aided Process Engineering*, Barcelona, Spain, 2005.
65. T.J. Courant. Dirac manifolds. *Trans. American Math. Soc. 319*, pages 631–661, 1990.
66. M. Dalsmo and A.J. van der Schaft. On representation and integrability of mathematical structures in energy conserving phyisical systems. *SIAM Journal of Control and Optimization*, 37(1), 1998.
67. M. De Leon, J. Cortes, D.M. De Diego, and S. Martinez. General symmetries in optimal control. *Reports on Mathematical Physics*, 53(1), 2004.
68. N. Diolaiti and C. Melchiorri. Tele-operation of a mobile robot through haptic feedback. In *HAVE 2002, IEEE Int. Workshop on Haptic Virtual Environments and Their Applications*, Ottawa, Canada, November 17-18 2002.
69. N. Diolaiti, C. Melchiorri, and S. Stramigioli. Contact impedance estimation for robotic systems. In *Proceedings of IEEE/RSJ International Conference on Intelligent Robots and Systems*, Sendai, Japan, September 2004.
70. N. Diolaiti, C. Melchiorri, and S. Stramigioli. Contact impedance estimation for robotic systems. *IEEE Transactions on Robotics*, 21(5):925–935, October 2005.
71. N. Diolaiti, G. Niemeyer, F. Barbagli, and J.K. Salisbury. Stability of haptic rendering: Discretization, quantization, time delay, and coulomb effects. *IEEE Transactions on Robotics*, 22(2):256–268, 2005.
72. B. A. Dubrovin, A.T. Fomenko, and S.P. Novikov. *Modern Geometry - Methods and Applications, Part I: The Geometry of Surfaces, Transformation Groups, and Fields*, volume 93. Springer-Verlag New York, Graduate texts in mathematics, ii edition, 1992.
73. B. A. Dubrovin, A.T. Fomenko, and S.P. Novikov. *Modern Geometry - Methods and Applications, Part II:The geometry and topology of manifolds*, volume 104. Springer-Verlag New York, Graduate texts in mathematics, ii edition, 1992.
74. J. Duffy. The fallacy of modern hybrid control theory that is based on "orthogonal complements" of twist and wrench space. *Journal of Robotic Systems*, 7(2), 1990.
75. V. Duindam. *Port-based modeling and Control for efficient bipedal walking robot.* PhD thesis, University of Twente, 2006.
76. V. Duindam and S. Stramigioli. Port-based asymptotic curve tracking for mechanical systems. *European Journal of Control*, 10(5):411–420, 2004.
77. V. Duindam and S. Stramigioli. Optimization of mass and stiffness distribution for efficient bipedal walking. In *Proceedings of IEEE/RSJ International Conference on Intelligent Robots and Systems*, Edmonton, Canada, August 2005.
78. V. Duindam and S. Stramigioli. Port-based control of a compass-gait bipedal robot. In *Proceedings of IFAC world Congress*, Prague, Czech Republic, July 2005.
79. D. Eberard and B. Maschke. An extension of port hamiltonian systems with boundary port variables to irreversible systems: the example of the heat conduction. In *Proceedings of Nonlinear Control Systems Design Symposium*, Stuttgart, Germany, 2004.

80. D. Eberard, B. Maschke, and A.J. van der Schaft. Port contact systems for irreversible thermodynamical systems. In *Proceedings of IEEE Conference on Decision and Control*, Seville, Spain, December 2005.
81. K.S. Eom, I.H. Suh, and B.-J. Yi. A design method of haptic interface controller considering transparency and robust stability. In *Proceedings of IEEE/RSJ International Conference on Intelligent Robots and Systems*, Lausanne, Switzerland, September 2002.
82. G. Escobar and H. Sira-Ramirez. A passivity based-sliding mode control approach for the regulation of power factors precompensators. In *Proceedings to the 37th Conference on Decision and Control*, December 1998.
83. G. Escobar, A.J. van der Schaft, and R. Ortega. A Hamiltonian viewpoint in the modeling of switching power converters. *Automatica*, 35(3), 1999.
84. X. Fan and M. Arcak. Delay robustness of a class of nonlinear systems and applications to communication networks. *IEEE Transactions on Automatic Control*, 51(1):139–144, January 2006.
85. C. Fantuzzi and C. Secchi. Energetic approach to parametric fault detection and isolation. In *Proceedings of American Control Conference*, Boston, Massachusetts, 2004.
86. C. Fantuzzi, C. Secchi, and A. Visioli. On the fault detection and isolation of industrial robot manipulators. In *Proceedings of the 7th International IFAC Symposium on Robot Control*, Wroclaw, Poland, September 2003.
87. A. Faraz, S. Payandeh, and A. Salvarinov. Design of haptic interface through stiffness modulation for endosurgery: Theory and experiments. In *Proceedings to IEEE International Conference on Robotics and Automation*, Leuven, Belgium, May 1998.
88. E.L. Faulring, K.M. Lynch, E. Colgate, and M.A. Peshkin. Haptic interaction with constrained dynamic systems. In *Proceedings of IEEE International Conference on Robotics and Automation*, Barcelona, Spain, April 2005.
89. R.L. Feller, C.K.L. Lau, D.P. Wagner, D.P. Perrin, and R.D. Howe. The effect of force feedback on remote palpation. In *Proceedings of IEEE International Conference on Robotics and Automation*, New Orleans, LA, USA, 2004.
90. W.R. Ferrell. Remote manipulation with transmission delay. *IEEE Transactions in Human Factors in Electronics*, HFE-6(1):24–32, September 1965.
91. W.R. Ferrell. Delayed force feedback. *IEEE Transactions in Human Factors in Electronics*, HFE-8:449–455, October 1966.
92. K.B. Fite, M. Goldfarb, and A. Rubio. Transparent telemanipulation in the presence of time delay. In *Proceedings of IEEE/ASME International Conference on Advanced Intelligent Mechatronics*, Kobe, Japan, July 2003.
93. K.B. Fite, L. Shao, and M. Goldfarb. Loop shaping for transparency and stability robustness in bilateral telemanipulation. *IEEE Transactions on Robotics and Automation*, 20(3):620–624, June 2004.
94. H. Flemmer, B. Eriksson, and J. Wikander. Control design for transparent teleoperators with model parameter variation. In *Proceedings of IEEE International Conference on Robotics and Automation*, Washington DC, USA, May 2002.
95. A. Fradkov. Exploring nonlinearity by feedback. *Physica D*, 128, 1999.
96. A. Fradkov. A Nonlinear Philosophy for Nonlinear Systems. In *Proceedings of IEEE Conference on Decision and Control*, Sydney, Australia, December 2000.

97. K. Fujimoto, K. Sakurama, and T. Sugie. Trajectory tracking control of port-controlled hamiltonian systems via generalized canonical transformations. *Automatica*, 2003.
98. E. Garcia-Canseco, A. Astolfi, and R. Ortega. Interconnection and damping assignment passivity-based control: towards a constructive procedure - part i and part ii. In *Proceedings of IEEE Conference on Decision and Control*, Paradise Island, Bahamas, December 2004.
99. C. Gaviria, E. Fossas, and R. Grino. Robust controller for a full-bridge rectifier using the ida approach and gssa modeling. *IEEE Transactions on Circuits and Systems*, 52(3):609–616, March 2005.
100. L. Gentili and A.J. van der Schaft. Regulation and input disturbance suppression for port-controlled hamiltonian systems. In *Proceedings of the 2nd Workshop on Lagrangian and Hamiltonian Methods for Nonlinear Control*, Seville, April 2003.
101. G. Gilardi and I. Sharf. Literature survey of contact dynamics modeling. *J. Mech. Mach. Theory*, 37(10):1213–1239, 2002.
102. B. Gillespie and M. Cutkosky. Stable user-specific haptic rendering of the virtual wall. In *Proceedings of the ASME International Mechanical Engineering Conference and Exposition*, 1996.
103. G. Golo, S. Stramigioli, and A.J. van der Schaft. Hamiltonian formulation of planar beams. In IFAC, editor, *Proceedings of the 2nd Workshop on Lagrangian and Hamiltonian Methods for Nonlinear Control*, pages 169–174, Sevilla, Spain, April 2003.
104. G. Golo, V. Talasila, and A.J. van der Schaft. Approximation of the telegrapher's equations. In *Proceedings of IEEE Conference on Decision and Control*, Las Vegas, USA, December 2002.
105. G. Golo, V. Talasila, A.J. van der Schaft, and B. Maschke. Hamiltonian discretization of boundary control systems. *Automatica*, 40(5), 2004.
106. G. Golo, V. Talasila, A.J. van der Schaft, and B.M. Maschke. Hamiltonian discretization of telegrapher's equations. draft, 2003.
107. G. Golo, A.J. van der Schaft, P.C. Breedveld, and B. Maschke. Hamiltonian formulation of bond graphs. In R. Johansson and A. Rantzer, editors, *Nonlinear and Hybrid Systems in Automotive Control*, pages 351–372. Springer-Verlag, 2003.
108. F. Gomez-Estern, R. Ortega, F.R. Rubio, and J. Aracil. Stabilization of a class of underactuated mechanical systems via total energy shaping. In *Proceedings of IEEE Conference on Decision and Control*, San Francisco, USA, December 2001.
109. O. Gonzalez. Time integration and discrete hamiltonian systems. *Journal of Nonlinear Science*, 1996.
110. W.H. Haddad, S.G. Nersesov, and V. Chellaboina. Energy-based control for hybrid port-controlled Hamiltonian systems. *Automatica*, 39(8), 2003.
111. W.M. Haddad, V.S. Chellaboina, and T. Rajpurohit. Dissipativity theory for nonnegative and compartmental dynamical systems with time delay. *IEEE Transactions on Automatic Control*, 49(5):747–751, May 2004.
112. B. Hannaford. A design framework for teleoperators with kinesthetic feedback. *IEEE Transactions on Robotics and Automation*, 5(4), 1989.
113. B. Hannaford. Stability and performance tradeoffs in bi-lateral telemanipulation. In *Proceedings of IEEE International Conference on Robotics and Automation*, Scottsdale, AZ, USA, May 1989.

114. B. Hannaford and J.-H. Ryu. Time domain passivity control of haptic interfaces. In *Proceedings of IEEE International Conference on Robotics and Automation*, San Francisco, CA, USA, May 2001.
115. B. Hannaford and J.-H. Ryu. Time domain passivity control of haptic interfaces. *IEEE Transactions on Robotics and Automation*, 18(1):1–10, February 2002.
116. K. Hashtrudi-Zaad and S.E. Salcudean. Transparency in time-delayed systems and the effect of local force feedback for transparent teleoperation. *IEEE Transactions on Robotics and Automation*, 18(1), February 2002.
117. I.N. Herstein. *Topics in Algebra*. John Wiley & Sons, 1975.
118. D.J. Hill. Dissipative nonlinear systems: Basic properties and stability analysis. In *Proceedings of IEEE Conference on Decision and Control*, Tucson, Arizona, December 1992.
119. S. Hirche and M. Buss. Passive position controlled telepresence systems with time delay. In *Proceedings of American Control Conference*, volume 1, Denver, Colorado, USA, June 2003.
120. S. Hirche and M. Buss. Study of teleoperation using realtime communication network emulation. In *Proceedings of IEEE/ASME International Conference on Advanced Intelligent Mechatronics*, Kobe, Japan, July 2003.
121. S. Hirche, B. Stanczyk, and M. Buss. Transparent Exploration of Remote Environments by Internet Telepresence. In *Proceedings of Int. Workshop on High-Fidelity Telepresence and Teleaction jointly with the conference HUMANOIDS*, Munich, Germany, 2003.
122. N. Hogan. Impedance control: An approach to manipulation: Part I – Theory, Part II – Implementation, Part III – Applications. *ASME Journal of Dynamic Systems, Measurment and Control*, 107, 1985.
123. N. Hogan. Controlling impedance at the man/machine. In *Proceedings IEEE International Conference on Robotics and Automation*, pages 1626–1631, 1989.
124. P.F. Hokayem and M.W. Spong. Bilateral teleoperation: An historical survey. *Automatica*, in press.
125. R. Horowitz and P. Li. Passive velocity field control (PVFC). part I. Geometry and robustntess. *IEEE Transactions on Automatic Control*, 46(9), 2001.
126. K.H. Hunt and F.R.E. Crossley. Coefficient of restitution interpreted as damping in vibroimpact. *ASME Journal of Applied Mechanics*, pages 440–445, June 1975.
127. D-Y Hwang and B. Hannaford. Modeling and stability analysis of a scaled telemanipulation system. In *Proceedings of IEE International Workshop on Robot and Human Communication*, 1994.
128. A. Isidori. *Nonlinear Control Systems*. Communication and Control Engineering Series. Springer-Verlag, 1995.
129. A. Isidori. *Nonlinear Control Systems II*. Communication and Control Engineering Series. Springer, December 1999. ISBN: 1852331887.
130. T. Itoh, K. Kosuge, and T. Fukuda. Human-machine cooperative telemanipulation wit motion and force scaling using task-oriented virtual tool dynamics. *IEEE Transactions on Robotics and Automation*, 16(5), 2003.
131. T. Iwasaki, S. Hara, and A.L. Fradkov. Time domain interpretations of frequency domain inequalities on (semi)finite ranges. *Systems and Control Letters*, 54(7):681–691, 2005.
132. B. Jacobs. *Objects and classes, co-algebraically*, chapter Object-Orientation with Parallelism and Persistence. Kluwer Acad. Publ., 1996.

133. B.P.F. Jacobs. Inheritance and cofree constructions. In *Proceedings of the European Conference on object-oriented programming*, 1996.
134. A.F. Jeandel, F. Boudard, Ph. Ravier, and A. Bushing. ULM: a modelling language. In *Proceedings of the CESA IMACS Multiconference*, pages 781–789, Lille, France, July 1996.
135. D. Jeltsema and J.M.A. Scherpen. A dual relation between port-Hamiltonian systems and the Brayton-Moser equations for nonlinear switched RLC circuits. *Automatica*, 39(6), 2003.
136. D. Jeltsema and J.M.A. Scherpen. Tuning of passivity-preserving controllers for switched-mode power converters. *IEEE Transactions on Automatic Control*, 49(8), 2004.
137. V. Jurdjevic and J.P. Quinn. Controllability and stability. *Journal of Differential Equations*, 28:381–389, 1978.
138. T. Kailath, A.H. Sayed, and B. Hassibi. *Linear Estimation.* Information and System Sciences. Prentice Hall, 2000.
139. K. Kaneko, H. Tokashiki, K. Tanie, and K. Komoriya. Impedance shaping based on force-feedback bilateral control in macro-micro teleoperation system. In *Proceedings of IEEE International Conference on Robotics and Automation*, Albuquerque, New Mexico, USA, April 1997.
140. D. Karnopp. Understanding multibody dynamics using bond graph representation. *Journal of the Franklin Institute*, 334B(4):631–642, 1997.
141. D.C. Karnopp, D.L. Margolis, and R.C. Rosenberg. *System Dynamics: A Unified Approach.* John Wiley & Sons Inc., second edition, 1990.
142. R. Kelly. Regulation of manipulators in generic task space: An energy shaping plus damping injection approach. *IEEE Transactions on Robotics and Automation*, 15(2), 1999.
143. Khalil. *Nonlinear Systems.* Pretience Hall, 1996.
144. V.L. Kharitonov and S-I Niculescu. On the stability of linear systems with uncertain delay. *IEEE Transactions on Automatic Control*, 48(1):127–132, January 2003.
145. W.S. Kim. Developments of new force reflecting control schemes and an application to a teleoperation training simulator. In *Proceedings of IEEE International Conference on Robotics and Automation*, pages 1412–1419, Nice, France, May 1992.
146. W.S. Kim and L-W. Stark. Cooperative control of visual displays for telemanipulation. In *Proceedings of IEEE International Conference on Robotics and Automation*, pages 1327–1332, 1989 1989.
147. Y-S. Kim and B. Hannaford. Some practical issues in time domain passivity control of haptic interfaces. In *Proceedings of IEEE/RSJ International Conference on Intelligent Robots and Systems*, Maui, Hawaii, October 2001.
148. K. Kosuge, T. Itoh, and T. Fukuda. Scaled telemanipulation with communication time delay. In *Proceedings of IEEE International Conference on Robotics and Automation*, Minneapolis, MN, USA, April 1996.
149. R. Krechetnikov and J.E. Marsden. On destabilizing effects of two fundamental non-conservative forces. *Physica D*, 214:25–32, 2006.
150. K.J. Kuchenbecker, J. Fiene, and G. Niemeyer. Event-based haptics and acceleration matching: Portraying and assessing the realism of contact. In *Proceedings of the First Joint Eurohaptics Conference and Symposium on Haptic Interfaces for Virtual Environment and Teleoperator Systems*, 2005.

151. K.J. Kuchenbecker and G. Niemeyer. Canceling induced master motion in force-reflectibg teleoperation. In *Proceedings of ASME Int. Mechanical Engineering Congress and Exposition*, Anaheim, CA, USA, November 2004.
152. G. Laferriere, G.J. Pappas, and S. Sastry. O-minimal hybrid systems. *Mathematics of Control, Signals and Systems*, 13:1–21, 2000.
153. C. Landi, A. Prati, and C. Secchi. Image analysis and automatic composition of ceramic mosaics. In *Proceedings of the VII Congress of the Italian societ for Industrial and Applied Mathematics*, Ragusa, Italy, May 2006.
154. M. Larsen, M. Jancovic, and P.V. Kokotovic. Coordinated passivation design. *Automatica*, 39(1), 2003.
155. D.A. Lawrence. Stability and transparency in bilateral teleoperation. *IEEE Transactions on Robotics and Automation*, 1993.
156. D.A. Lawrence, L.Y. Pao, A.M. Dougherty, M.A. Salada, and Y. Pavlou. Rate-hardness: A new performance metric for haptic interfaces. *IEEE Transactions on Robotics and Automation*, 16(4), 2000.
157. C.D. Lee, D.A. Lawrence, and L.Y. Pao. Isotropic force control for haptics interfaces. *Control Engineering Practice*, 12(11), November 2004.
158. D. Lee and P.I. Li. Passive bilateral feedforward control of linear dynamically similar teleoperated manipulators. *IEEE Transactions on Robotics and Automation*, 2003.
159. D. Lee and M.W. Spong. Passive bilateral control of teleoperators under constant time-delay. In *Proceedings of IFAC world Congress*, Prague, Czech Republic, July 2005.
160. D. Lee and M.W. Spong. Passive bilateral control of teleoperators under constant time-delay. *submitted to IEEE Transactions on Robotics*, 2005.
161. E.A. Lee. Overview of the Ptolemy project. Technical report, University of California, Berkeley, USA, July 2003.
162. E.A. Lee and A. Sangiovanni-Vicentelli. A framework for comparing models of computation. *IEEE Transactions on Computer Aided Design*, 17(12), 1998.
163. Singha Leeraphan, Thavida Maneewarn, and Djitt Laowattana. Stable adaptive bilateral control of trasparent teleoperation through time-varying delay. In *Proceedings to IEEE/RSJ International Conference on Intelligent Robots and Systems*, EPFL, Lausanne, Switzerland, October 2002.
164. A. Lewis, C.T. Abdallah, and D.M. Dawson. *Control of Robot Manipulator.* Macmillan Publ. Co., 1993.
165. F.B. Llewellyn. Some fundamental properties of transmission systems. *Proceedings of IRE*, 40:271–283, 1952.
166. W. Lohmiller and J.-J. Slotine. On contraction analysis for non-linear systems. *Automatica*, 34(6), 1998.
167. W. Lohmiller and J.-J. Slotine. Control system design for mechanical systems using contraction theory. *IEEE Transactions on Automatic Control*, 45(5), May 2000.
168. J. Loncaric. *Geometrical Analysis of Compliant Mechanisms in Robotics.* PhD thesis, Harvard University, Cambridge, MA, April 1985.
169. R.G. Longoria. Wave-scattering formalism for multiport energetic systems. *Journal of the Franklin Institute*, 333(4):539–564, 1996.
170. R. Lozano, B. Brogliato, O. Egeland, and B. Maschke. *Dissipative Systems, Analysis and Control.* Communication and Control Engineering. Springer, London, 2000.

171. R. Lozano, N. Chopra, and M.W. Spong. Passivation of force reflecting bilateral teleoperators with time varying delay. In *Proceedings to Mechatronics Forum International Conference*, Enschede, The Netherlands, June 2002.
172. S. Mac Lane. *Categories for the working mathematician.* Springer-Verlag, 1971.
173. A. Macchelli. *Port Hamiltonian systems. A unified approach for modeling and control finite and infinite dimensional physical systems.* PhD thesis, University of Bologna, 2004.
174. A. Macchelli and C. Melchiorri. Modeling and control of the timoshenko beam. the distributed port hamiltonian approach. *SIAM Journal On Control and Optimization (SICON)*, 43(2):743–767, 2004.
175. A. Macchelli and C. Melchiorri. Control by interconnection of mixed port hamiltonian systems. *IEEE Transactions on Automatic Control*, 50(11):289–293, November 2005.
176. A. Macchelli, C. Melchiorri, and L. Bassi. Port-based modelling and control of the mindlin plate. In *Proceedings of IEEE Conference on Decision and Control*, Seville, Spain, December 2005.
177. A. Macchelli, C. Melchiorri, C. Secchi, and C. Fantuzzi. A variable structure approach to energy shaping. In *Proceedings of the European Control Conference*, Cambridge, UK, September 2003.
178. A. Macchelli, C. Melchiorri, S. Stramigioli, A.J. van der Schaft, and C. Secchi. On the constrained dynamics of port-Hamiltonian systems: Theory and applications. *submitted to Automatica*, 2003.
179. A. Macchelli, S. Stramigioli, A.J. van der Schaft, and M. Melchiorri. Scattering for infinite dimensional port-hamiltonian systems. In *Proceedings of IEEE Conference on Decision and Control*, Las Vegas, Nevada, December 2002.
180. S. Mahapatra and M. Zefran. Stable haptic interaction with switched virtual environments. In *Proceedings of IEEE International Conference on Robotics and Automation*, Taipei, Taiwan, ROC, September 2003.
181. M. Mahvash and V. Hayward. High-fidelity passive force reflecting virtual environments. *IEEE Transactions on Robotics*, 21(1), February 2005.
182. D.H.S. Maithripala, J.M. Berg, and W.P. Dayawansa. Almost global tracking of simple mechanical systems on a general class of Lie groups. *IEEE Transactions on Automatic Control*, 51(1):216–225, January 2006.
183. S. Manorotkul and H. Hashimoto. Virtual impedance with position error correction for teleoperator with time delay. In *Proceedings of the 4th International Workshop on Advanced Motion Control*, March 1996.
184. D.W. Marhefka and D.E. Orin. A compliant contact model with nonlinear damping for simulation of robotic systems. *IEEE Transactions on systems, man and cybernetics - Part A: Systems and Humans*, 29(6), November 1999.
185. J.E. Marsden, R. Abraham, and T. Ratiu. *Manifolds, Tensor Analisys and Applications.* Springer-Verlag, 1988.
186. J.E. Marsden and T.S. Ratiu. *Introduction to Mechanics and Symmetry.* Number 17 in Texts in Applied Mathematics. Springer-Verlag New York, 2nd edition, 1999.
187. Ph. Martin, R.M. Murray, and P. Rouchon. Flat systems. Lectures given at the Summer school on mathematical control theory, Trieste, Italy, September 2001.
188. B. Maschke, R. Ortega, and A.J. van der Schaft. Energy-based lyapunov functions for forced Hamiltonian systems with dissipation. In *Proceedings to the 37th Conference on Decision and Control (CDC)*, 1998.

189. B. Maschke and A.J. van der Schaft. Port controlled hamiltonian systems: modeling origins and system theoretic properties. In *Proceedings of the third Conference on nonlinear control systems (NOLCOS)*, 1992.
190. B. Maschke and A.J. van der Schaft. Port controlled hamiltonian representation of distributed parameter sytems. In *Workshop on modeling and Control of Lagrangian and Hamiltonian Systems*, 2000.
191. B. Maschke, A.J. van der Schaft, and P. Breedveld. An intrinsic hamiltonian formulation of network dynamics: Non-standard poisson structures and gyrators. *Journal of the Franklin institute*, 1992.
192. B.M. Maschke and A.J. van der Schaft. Hamiltonian systems, pseudo poisson brackets and their scattering representation for physical systems. In *Proceedings of the International Symposium on Motion and Vibration Control, 17th ASME Biennal Conference on Mechanical Vibration and Noise*, Las Vegas, Nevada, September 1999.
193. B.M. Maschke, A.J. van der Schaft, and C. Bidard. An extension of scattering variables to spatial mechanisms. In *Proceedings of 3rd IMACS Conference on Mathematical Modeling*, Vienna, Austria, February 2000.
194. S.G. Massaquoi and J.-J. Slotine. The intermediate cerebellum may function as a wave-variable processor. *Neuroscience Letters*, 215:60–64, 1996.
195. S.E. Mattsson and M.; Andersson. The ideas behind omola. In *Proceedings of the IEEE Symposium on Computer-Aided Control System Design*, pages 23–29, March 1992.
196. R. I. McLachlan, G.R.W. Quispel, and N. Robidoux. Unified approach to Hamiltonian systems, Poisson systems, gradient systems with lyapunov functions or first integrals. *Physical Review Letters*, 81(12), 1998.
197. C. Melchiorri and A. Eusebi. *Int. Summer School on Modeling and Control of Mechanisms and Robot*, chapter Telemanipulation: System Aspect and Control Issues. World Scientific, 1996.
198. W. Michiels, V. van Assche, and S.-I. Niculescu. Stabilization of time-delay systems with a controlled time-varying delay and applications. *IEEE Transactions on Automatic Control*, 50(4):493–504, 2005.
199. B.E. Miller, J.E. Colgate, and R.A. Freeman. Computational delay and free mode environment design for haptic display. In *Proceedings of the Internationa Mechanical Engineering Congress and Exhibition*, Nashville, Tennessee, 1999.
200. B.E. Miller, J.E. Colgate, and R.A. Freeman. Passive implementation for a class of static nonlinear environments in haptic display. In *Proceedings of IEEE International Conference on Robotics and Automation*, Detroit, Michigan, May 1999.
201. B.E. Miller, J.E. Colgate, and R.A. Freeman. Environment delay in haptic systems. In *Proceedings to IEEE International Conference on Robotics and Automation*, San Francisco, USA, April 2000.
202. B.E. Miller, J.E. Colgate, and R.A. Freeman. Guaranteed stability of haptic systems with nonlinear virtual environnments. *IEEE Transactions on Robotics and Automation*, 16(6):712–719, December 2000.
203. B.E. Miller, J.E. Colgate, and R.A. Freeman. On the role of dissipation in haptic systems. *IEEE Transactions on Robotics*, 20(4), August 2004.
204. J.W. Milnor and J.D. Stasheff. *Characteristic classes.* Princeton University Press, 1974.
205. I. Miroshnik. Attractors and partial stability of nonlinear dynamical systems. *Automatica*, 40(3), 2004.

206. F. Morabito, A.R. Teel, and L. Zaccarian. Nonlinear antiwindup applied to euler-lagrange systems. *IEEE Transactions on Automatic Control*, 20(3), June 2004.
207. L. Moreau and D. Aeyels. A novel variational method for deriving lagrangian and hamiltonian models of inductor-capacitor circuits. *SIAM Review*, 46(1):59–84, 2004.
208. P.J. Mosterman, J. Sztipanovits, and S. Engell. Computer-automated multiparadigm modeling in control systems technology. *IEEE Transactions on Control Systems Technology*, 12(2):223–234, March 2004.
209. P.J. Moylan. A connective stability result for interconnected passive systems. *IEEE Transactions on Automatic Control*, 25(4), August 1980.
210. S. Munir and W.J. Book. Wave-based teleoperation with prediction. In *Proceedings of American Control Conference*, Arlington, VA, USA, June 2001.
211. S. Munir and W.J. Book. Internet-based teleoperation using wave variables with prediction. *IEEE Transactions on Mechatronics*, 7(2), June 2002.
212. R.M. Murray, Z. Li, and S.S. Sastry. *A mathematical introduction to robotic manipulation*. CRC press, 1994.
213. E.M. Navarro-Lopez and E. Fossas-Colet. Feedback passivity of nonlinear discrete-time systems with direct input-output link. *Automatica*, 8(40), 2004.
214. G. Niemeyer. *Using wave variables in time delayed force reflecting teleoperation*. PhD thesis, Massachusetts Institute of Technology, 1996.
215. G. Niemeyer and J. Slotine. Stable adaptive teleoperation. *IEEE Journal of Oceanic Engineering*, 16(1):152–162, 1991.
216. G. Niemeyer and J.J.E. Slotine. Using wave variables for system analysis and robot control. In *Proceedings to IEEE International Conference on Robotics and Automation*, Albuquerque, New Mexico, April 1997.
217. G. Niemeyer and J.J.E. Slotine. Towards force-reflecting teleoperation over the internet. In *Proceedings to International Conference on Robotics and Automation*, Leuven, Belgium, May 1998.
218. J.-J. Niemeyer, G. ans Slotine. Telemanipulation with Time Delays. *International Journal of Robotics Research*, 23(9):873–890, September 2004.
219. H. Nijmeijer and A.J. van der Schaft. *Nonlinear dynamical control systems*. Springer-Verlag, 1991.
220. J.T. Nolin, P.M. Stemniski, and A.M. Okamura. Activation cues and force scaling methods for virtual fixtures. In *Proceedings of the 11th Symposium on Haptic Interfaces for Virtual Environment and Teleoperator Systems*, Chicago, IL, USA, 2003.
221. P.J. Olver. *Applications of Lie Groups to Differential Equations*. Springer, Berlin, second edition, 1993.
222. J.-P. Ortega and V. Planas-Biesla. Dynamics on Leibniz manifolds. *Journal of Geometry and Physics*, 52(1), 2004.
223. R. Ortega, Z.P. Jang, and D.J. Hill. Passivity-based control of nonlinear systems: A tutorial. In *Proceedings of American Control Conference*, Albuquerque, New Mexico, June 1997.
224. R. Ortega, D. Jeltsema, and J.M.A. Scherpen. Power shaping: A new paradigm for stabilization of nonlinear rlc circuits. In *Proceedings to the IFAC World Congress*, Barcelona, Spain, July 2002.

225. R. Ortega, A. Loria, R. Kelly, and L. Praly. On passivity-based output feedback global stabilization of euler-lagrange systems. In *Proceedings of IEEE Conference on Decision and Control*, Lake Buena Vista, Florida, December 1994.
226. R. Ortega, A. Loria, P.J. Nicklasson, and H. Sira-Ramirez. *Passivity-based Control of Euler-Lagrange Systems.* Springer-Verlag, London, UK, 1998.
227. R. Ortega, A.J. van der Schaft, I. Mareels, and B. Maschke. Putting energy back in control. *IEEE Control Systems Magazine*, 21(2), April 2001.
228. R. Ortega, A.J. van der Schaft, B. Maschke, and G. Escobar. Energy-shaping of port-controlled hamiltonian systems by interconnection. In *Proceedings of IEEE Conference on Decision and Control*, Phoenix, Arizona, December 1999.
229. R. Ortega, A.J. van der Schaft, B. Maschke, and G. Escobar. Interconnection and damping assignment passivity-based control of port-controlled hamiltonian systems. *Automatica*, 38, 2002.
230. G.J. Pappas. Bisimilar linear systems. *Automatica*, 39:2035–2047, 2003.
231. G.J. Pappas, G. Laferriere, and S. Sastry. Hierarchically consistent control systems. *IEEE Transactions on Automatic Control*, 45(6):1144–1160, 2000.
232. G.J. Pappas and S. Simic. Consistent abstractions of affine control systems. *IEEE Transactions on Automatic Control*, 47(5):745–756, 2002.
233. H.-S. Park, P.H. Chang, and J.H. Kim. Significance of causality analysis for transparent teleoperation. In *Proceedings of IEEE International Conference on Robotics and Automation*, pages 709–714, Barcelona, Spain, April 2005.
234. J. Park and W.-K. Chung. Geometric integration on euclidean group with application to articulated multibody systems. *IEEE Transactions on Robotics*, 21(5):850–863, October 2005.
235. J.H. Park and H.C. Cho. Sliding mode controller for bilateral teleoperation with varying time delay. In *Proceedings of IEEE/ASME International Conference on Advanced Intelligent Mechatronics*, pages 311–316, Atlanta, GA, USA, September 1999.
236. S. Payandeh and Z. Stanisic. On application of virtual fixtures as an aid for telemanipulation and training. In *Proceedings of the Symposium on Haptic Interfaces for Virtual Environment and Teleoperator Systems*, 2002.
237. H.M. Paynter. *Analysis and Design of Engineering Systems.* M.I.T. Press, 1961.
238. M. Perez, R. Ortega, and J. Espinosa. Passivity–based pi control of switched power converters. In *Proceedings of European Control Conference*, Cambridge, UK, October 2003.
239. C. Perkins, O. Hodson, and V. Hardman. A survey of packet loss recovery techniques for streaming audio. *IEEE Network*, 12(5), 1998.
240. V. Petrovic, R. Ortega, and A.M. Stankovic. Interconnection and damping assignment approach to control of pm synchronous motors. *IEEE Transactions on Control Systems Technology*, 9(6):811–820, November 2001.
241. P. Piela. *ASCEND: An Object-Oriented Computer Environment for Modeling and Analysis.* PhD thesis, Carnegie Mellon University, 1989.
242. G. Pola, A.J. van der Schaft, and M.D. Di Benedetto. Bisimulation theory for switching linear systems. In *Proceedings of IEEE Conference on Decision and Control*, Paradise Island, Bahamas, 2004.
243. J.W. Poldermann and J.C. Willems. *Introduction to Mathematical Systems Theory - A Behavioral Approach.* Number 26 in Texts in Applied Mathematics. Springer-Verlag New York, 1998.

244. I.G. Polushin and H.J. Marquez. Boundedness properties of nonlinear quasi-dissipative systems. *IEEE Transactions on Automatic Control*, 49(12), December 2004.

245. P.A. Prokopiu, S.G. Tzafestas, and W.S. HArwin. Towards variable-time-delays-robust telemanipulation through master state prediction. In *Proceedings of IEEE/ASME International Conference on Advanced Intelligent Mechatronics*, Atlanta, GA, USA, September 1999.

246. B. Recht and R. D'Andrea. Distributed control of systems over discrete groups. *IEEE Transactions on Automatic Control*, 49(9), September 2004.

247. H. Rodriguez, A.J. van der Schaft, and R. Ortega. On stabilization of nonlinear distributed parameter port-controlled hamiltonian systems via energy shaping. In *Proceedings of IEEE Conference on Decision and Control*, Orlando, Florida, December 2001.

248. l.B. Rosenberg. Virtual fixtures: Perceptual tools for telerobotic manipulation. In *Proceedings of the Virtual Reality Annual International Symposium,*, 1993.

249. J-H Ryu and B. Hannaford. Stable teleoperation with time-domain passivity control. *IEEE Transactions on Robotics and Automation*, 20(2), 2004.

250. J-H Ryu, B. Hannaford, C. Preusche, and G. Hirzinger. Time domain passivity control with refrence energy behavior. In *Proceedings of IEEE/RSJ International Conference on Intelligent Robots and Systems*, pages 2932–2937, Las Vegas, NV, October 2003.

251. J.-H. Ryu, J.H. Kim, D.-S. Kwon, and B. Hannaford. A simulation/experimental study of the noisy behavior of the time domain passivity controller for haptic interfaces. In *Proceedings of IEEE International Conference on Robotics and Automation*, Barcelona, Spain, April 2005.

252. J.-H. Ryu, Y.S. Kim, and B. Hannaford. Sampled- and continuous-time passivity and stability of virtual environments. *IEEE Transactions on Robotics*, 20(4):772–776, August 2004.

253. J-H. Ryu, D-S. Kwon, and B. Hannaford. Stability guaranteed control: Time domain passivity approach. In *Proceedings to IEEE/RSJ Conference on Intelligent Robots and Systems*, EPFL, Lausanne, Switzerland, October 2002.

254. J-H. Ryu, S.K. Yoon, and B. Hannaford. Sampled and continuous time passivity and stability of virtual environments. In *IEEE International Conference on Roboticss and Automation*, Taipei, Taiwan, September 2003.

255. S.E. Salcudean and T. Vlaar. On the emulation of stiff walls and static friction with a magnetically levitated input-output device. *ASME Journal of Dynamic Systems, Measurement and Control*, 119, March 1997.

256. C. Secchi. *Interactive Robotic Interfaces: a port-Hamiltonian Approach.* PhD thesis, University of Modena and Reggio Emilia, 2004. available at http://www.dismi.unimore.it/download/thesis.pdf.

257. C. Secchi, C. Fantuzzi, and M. Bonfé. On the use of uml for modeling physical systems. In *Proceedings of IEEE International Conference on Robotics and Automation*, Barcelona, Spain, April 2005.

258. C. Secchi, C. Fantuzzi, and M. Bonfé. Unified modeling of control software and physical plants. In *Proceedings of IFAC world Congress*, Prague, Czech Republic, July 2005.

259. C. Secchi, C. Fantuzzi, and A. Gianotti. Control of an industrial manipulator using rtai linux. In *Fifth Real-time Linux Workshop*, Valencia, Spain, November 2003.

260. C. Secchi, Bonfé M., and C. Fantuzzi. On the use of uml for modeling mechatronic systems. *IEEE Transactions on Automation Science and Engineering*, 2006. In press.
261. C. Secchi, S. Stramigioli, and C. Fantuzzi. Dealing with unreliabilities in digital passive geometric telemanipulation. In *Proceedings of IEEE/RSJ International Conference on Intelligent Robots and Systems*, Las Vegas, NV, October 2003.
262. C. Secchi, S. Stramigioli, and C. Fantuzzi. Delayed virtual environments: a port-hamiltonian approach. In *Proceedings of IEEE/RSJ International Conference on Intelligent Robots and Systems*, Las Vegas, NV, October 2003.
263. C. Secchi, S. Stramigioli, and C. Fantuzzi. Digital passive geometric telemanipulation. In *Proceedings of IEEE International Conference on Robotics and Automation*, Taipei, Taiwan, September 2003.
264. C. Secchi, S. Stramigioli, and C. Fantuzzi. Power scaling in port-hamiltonian based bilateral telemanipulation. In *Proceedings of IEEE/RSJ International Conference on Intelligent Robots and Systems*, Edmonton, Canada, August 2005.
265. C. Secchi, S. Stramigioli, and C. Fantuzzi. The problem of packets loss in scaled digital port-hamiltonian based bilateral telemanipulation. In *Proceedings of IEEE Conference on Control Applications*, Toronto, Canada, August 2005.
266. C. Secchi, S. Stramigioli, and C. Fantuzzi. Transparency in port-hamiltonian based telemanipulation. In *Proceedings of IEEE/RSJ International Conference on Intelligent Robots and Systems*, Edmonton, Canada, August 2005.
267. C. Secchi, S. Stramigioli, and C. Fantuzzi. Intrinsically passive force scaling in haptic interfaces. In *Proceedings of IEEE/RSJ International Conference on Intelligent Robots and Systems*, Beijing, China, October 2006.
268. C. Secchi, S. Stramigioli, and C. Fantuzzi. Kinematic compensation in port-hamiltonian telemanipulation. In *Proceedings of IFAC Workshop on Lagrangian and Hamiltonian Methods for Nonlinear Control*, Nagoya, Japan, July 2006.
269. C. Secchi, S. Stramigioli, and C. Fantuzzi. Position drift compensation in port-hamiltonian based telemanipulation. In *Proceedings of IEEE/RSJ International Conference on Intelligent Robots and Systems*, Beijing, China, 2006.
270. C. Secchi, S. Stramigioli, and C. Fantuzzi. Variable delay in scaled port-hamiltonian telemanipulation. In *Proceedings of the IFAC Symposium on Robot Control*, Bologna, Italy, September 2006.
271. C. Secchi, S. Stramigioli, and C. Melchiorri. Geometric grasping and telemanipulation. In *Proceedings to IEEE Conference on Intelligent Robotic Systems*, Maui, Hawaii, USA, October 2001.
272. J.M. Selig. *Geometric Methods in Robotics*. Monographs in Computer Sciences. Springer Verlag, 1996. ISBN 0-387-94728-0.
273. T.B. Sheridan. Space teleoperation through time delay: review and prognosis. *IEEE Transactions on Robotics and Automation*, 9(5):592–606, 1993.
274. M. Shi, T. Gang, S. Graves, and J.H. Downs. Positive realness and tracking of teleoperation systems. In *Proceedings of IEEE Conference on Decision and Control*, Tampa, FL, USA, December 1998.
275. H. Shim, J.H. Seo, and A.R. Teel. Nonlinear observer design via passivation of error dynamics. *Automatica*, 39(5), 2003.
276. A.S. Shiriaev, H. Ludvigsen, and O. Egeland. Swinging up the spherical pendulum via stabilization of its first integrals. *Automatica*, 40(1):73–85, 2004.

277. S. Shishkin, R. Ortega, D. Hill, and A. Loria. On output feedback stabilization of euler-lagrange systems with nondissipative forces. In *Proceedings of IEEE Conference on Decision and Control*, New Orleans, Louisiana, December 1995.
278. H. Sira-Ramirez. Differential methods in variable-structure control. *International Journal of Control*, 48(4), 1988.
279. H. Sira-Ramirez. Nonlinear variable structure systems in sliding mode: the general case. *IEEE Transactions on Automatic Control*, 1989.
280. H. Sira-Ramirez. A general canonical form for sliding mode control of nonlinear systems. In *Proceedings to the European Control Conference*, 1999.
281. H. Sira-Ramirez, G. Escobar, and R. Ortega. On passivity-based sliding mode control of switched dc-to-dc power converters. In *Proceedings of the 35th Conference on Decision and Control*, December 1996.
282. S. Sirouspour. Modeling and control of cooperative teleoperation systems. *IEEE Transactions on Robotics*, 2005.
283. J.-J. Slotine and W. Li. *Applied Nonlinear Control.* Prentice Hall, 1991.
284. E.D. Sontag. *Mathematical Control Theory: Deterministic Finite Dimensional Systems.* Texts in applied mathematics. Springer, 1998.
285. E.D. Sontag. Adaptation and regulation with signal detection implies internal model. *Systems and Control Letters*, 50, 2003.
286. E.D. Sontag. Passivity gains and the "secant condition" for stability. *Systems and Control Letters*, 55:177–183, 2006.
287. J.E. Speich and M. Goldfarb. Implementation of loop-shaping compensators to increase the transparency bandwidth of a scaled telemanipulation system. In *Proceedings of IEEE International Conference on Robotics and Automation*, Washington, DC, USA, May 2002.
288. S. Stramigioli. Creating artificial damping by means of damping injection. In K.Danai (Ed.), editor, *Proceedings of the ASME Dynamic Systems and Control Division*, volume DSC.58, 1996.
289. S. Stramigioli. *From Differentiable Manifolds to Interactive Robot Control.* PhD thesis, Delft University of Technology, Delft, The Netherlands, 1998.
290. S. Stramigioli. *Modeling and IPC Control of Interactive Mechanical Systems: a coordinate free approach.* LNCIS. Springer, London, 2001.
291. S. Stramigioli. About the use of port concepts for passive geometric telemanipulation with varying time delays. In *Proceedings to Mechatronics Forum International Conference*, Enschede, The Netherlands, June 2002.
292. S. Stramigioli, G. Blankenstein, V. Duindam, H. Bruyninckx, and C. Melchiorri. Power port concepts in robotics. Tutorial of Internationcal Conference on Robotics and Automation, September 2003.
293. S. Stramigioli and V. Duindam. Variable spatial springs for robot control applications. In *IROS*, Maui, Hawaii, October 2001.
294. S. Stramigioli and V. Duindam. Port-based modeling of spatial visco-elastic contacts. *European Journal of Control*, 10:510–519, 2004.
295. S. Stramigioli, E.D. Fasse, and J.C. Willems. A rigorous framework for interactive robot control. *International Journal of Control*, 75(11), 2002.
296. S. Stramigioli, B. Maschke, and A.J. van der Schaft. Passive output feedback and port interconnection. In *Symposium on Nonlinear Control Systems and Design*, 1998.
297. S. Stramigioli, C. Secchi, A.J. van der Schaft, and C. Fantuzzi. A novel theory for sample data system passivity. In *Proceedings to IEEE Conference on Intelligent Robotic Systems*, 2002.

298. S. Stramigioli, C. Secchi, A.J. van der Schaft, and C. Fantuzzi. Sampled data systems passivity and discrete port-hamiltonian systems. *IEEE Transactions on Robotics*, 21(4):574–587, 2005.
299. S. Stramigioli, A. van der Schaft, B. Maschke, and C. Melchiorri. Geometric scattering in robotic telemanipulation. *IEEE Transactions on Robotics and Automation*, 18(4), 2002.
300. S. Stramigioli, A. van der Schaft, B. Maschke, and C. Melchiorri. Geometric scattering in robotic telemanipulation. *IEEE Transactions on Robotics and Automation*, 18(4), 2002.
301. S. Stramigioli, A.J. van der Schaft, B. Maschke, S. Andreotti, and C. Melchiorri. Geometric scattering in telemanipulation of port controlled hamiltonian systems. In *Proceedings of the 39th IEEE Conference on Decision and Control*, Sidney, Australia, December 2000.
302. H.J. Sussmann and J.C. Willems. 300 years of optimal control: from the brachystochrone to the maximum principle. *Control Systems Magazine*, 17(3):32–44, 1997.
303. P. Tabuada and G.J. Pappas. From nonlinear to hamiltonian via feedback. In *Proceedings of IEEE Conference on Decision and Control*, Las Vegas, Nevada, December 2002.
304. P. Tabuada and G.J. Pappas. Abstractions of hamiltonian control systems. *Automatica*, 2003.
305. G. Takegaki and S. Arimoto. A new feedback method for dynamic control of manipulators. *ASME Journal of Dynamic Systems Measurement and Control*, 102, 1981.
306. M. Takegaki and S. Arimoto. A new feedback method for dynamic control of manipulators. *ASME Journal of Dynamic Systems, Measurement and Control*, 102:119–125, 1981.
307. K. Takep and K. Kosuge. Implementation of the micro-macro teleoperation system without using slave-side force sensors. In *Proceedings of IEEE International Conference on Robotics and Automation*, Albuquerque, New Mexico, April 1997.
308. V. Talasila, J. Clemente-Gallardo, and A.J. van der Schaft. Discrete port hamiltonian systems,. *Systems and Control Letters*, 55:478–486, 2006.
309. N. A. Tanner and G. Niemeyer. Online tuning of wave impedance in teleoperation. In *Proceedings of the IEEE Conference on Robotics, Automation and Mechatronics*, Singapore, December 2004.
310. N.A. Tanner and G. Niemeyer. Practical limitations of wave variable controllers in teleoperation. In *Proceedings of the IEEE Conference on Robotics, Automation and Mechatronics*, Singapore, December 2004.
311. N.A. Tanner and G. Niemeyer. Improving perception in time delayed teleoperation. In *Proceedings of IEEE International Conference on Robotics and Automation*, Barcelona, Spain, April 2005.
312. M. Tiller, editor. *Introduction to physical modeling with Modelica*. Kluwer Academic Publisher, 2001.
313. H.L. Trentelman. On flat systems behaviors and observable image representations. *Systems and Control Letters*, 51(1), January 2004.
314. T. Tsuji and K. Ohnishi. Position/force scaling of function-based bilateral control system. In *Proceedings of the IEEE International Conference on Industrial Technology*, 2004.

315. J. Ueda and T. Yoshikawa. Force-reflecting bilateral teleoperation with time delay by signal filtering. *IEEE Transactions on Robotics and Automation*, 20(3), June 2004.
316. A.C. Umarikar and L. Umanand. Modelling of switching systems in bond graphs using the concept of switched power junctions. *Journal of the Franklin Institute*, 342(2), March 2005.
317. V.I. Utkin. *Sliding regimes in the theory of variable structure system*. MIR Editor, Moscow, Russia, 1978.
318. A.J. van der Schaft. Implicit Hamiltonian systems with symmetry. *Reports on mathematical physics*, 41(2), 1998.
319. A.J. van der Schaft. *L_2-Gain and Passivity Techniques in Nonlinear Control*. Communication and Control Engineering. Springer Verlag, 2000.
320. A.J. van der Schaft. Equivalence of Dynamical Systems by Bisimulation. *IEEE Transactions on Automatic Control*, 49(12):2160–2172, 2004.
321. A.J. van der Schaft and J. Cervera. Composition of dirac structures and control of port-hamiltonian systems. In *Proceedings of the 15th International Symposium on Mathematical Theory of Networks and Systems*, South Bend, USA, 2002.
322. A.J. van der Schaft, M. Dalsmo, and B.M. Maschke. Mathematical structure in the network represenation of energy-conserving physical systems. In *Proceedings of IEEE Conference on Decision and Control*, Kobe, Japan, December 1996.
323. A.J. van der Schaft and B.M. Maschke. The Hamiltonian formulation of energy conserving physical systems with external ports. *Archiv für Elektronik und Übertragungstechnik*, 49(5/6):362–371, 1995.
324. A.J. van der Schaft and B.M. Maschke. Fluid dynamical systems as hamiltonian boundary control systems. In *Proceedings of IEEE Conference on Decision and Control*, Orlando, Florida, 2001 2001.
325. A.J. van der Schaft and B.M. Maschke. Hamiltonian formulation of distributed-parameter systems with boundary energy flow. *Journal of Geometry and Physics*, 42(1-2), May 2002.
326. J. Vertut and P. Coiffet. *Robot Technology, Volume 3A: Teleoperations and Robotics: Evolution and Development*. Englewood Cliffs, NJ: Prentice-Hall, 1986□.
327. Y. Wang, C. Li, and D. Cheng. Generalized Hamiltonian realization of time-invariant nonlinear systems. *Automatica*, 39(8), 2003.
328. D. Whitney. Force feedback control of manipulator fine motions. *ASME Journal of Dynamic Systems, Measurement and Control*, pages 91–97, 1977.
329. J.C. Willems. Dissipative dynamical systems, part i: General theory. *Arch. Rat. Mech. An.*, 45, 1972.
330. J.C. Willems. Paradigms and puzzles in the theory of dynamical systems. *IEEE Transactions on Automatic Control*, 36(3), 1991.
331. J.C. Willems. On Interconnections, Control and Feedback. *IEEE Transactions on Automatic Control*, 42(3), 1997.
332. G. Winskel and M. Nielsen. *Handbook of Logic and Foundations of Theoretical Computer Science*, chapter Models for Concurrency. Oxford University Press, 1994.
333. S. Yamakawa, A. Koji, and H. Fujimoto. Conditions of force scaling methods based on human perception abilities for displaying time-variant force. In *Proceedings of the SICE annual conference*, August 2004.

334. X. Ye, M. Q-H. Meng, P.X. Liu, and G. Li. Statistical analysis and prediction of round trip delay for internet-based teleoperation. In *Proceedings to IEEE/RSJ Conference on Intelligent Robots and Systems*, EPFL, Lausanne, Switzerland, October 2002.
335. P.-Y. Yen. Maximization of the stability allowable region of the impedance scaling for telemanipulators. In *Proceedings of the IEEE International Conference on Networking, Sensing & Control*, Taipei, Taiwan, March 2004.
336. Y. Yokokohji, T. Imaida, and T. Yoshikawa. Bilateral teleoperation under time-varying communication delay. In *Proceedings of IEEE/RSJ International Conference on Intelligent Robots and Systems*, 1999.
337. Y. Yokokohji, T. Imaida, and T. Yoshikawa. Bilateral control with energy balance monitoring under time-varying communication delay. In *Proceedings of IEEE International Conference on Robotics and Automation*, San Francisco, California, April 2000.
338. Y. Yokokohji, T. Tsujioka, and T. Yoshikawa. Bilateral control with time-varying delay including communication blackout. In *Proceedings of the 10th Symposyum on Haptic Interfaces for Virtual Environments and Teleoperator Systems*, 2002.
339. Y. Yokokohji and T. Yoshikawa. Bilateral control of master-slave manipulators for ideal kinesthetic coupling. In *Proceedings of IEEE International Conference on Robotics and Automation*, Nice, France, May 1992.
340. A. Zeid and C.-C. Chung. Bond graph modeling of multibody systems: a library of three-dimensional joints. *Journal of the Franklin Institute*, 329(4):605–636, 1992.
341. L. Zhihu, W. Jingcheng, and S. Huihe. Delay-dependent dissipative control for linear time-delay systems. *Journal of the Franklin Institute*, 339:529–542, 2002.
342. M. Zhu and S.E. Salcudean. Achieving transparency for teleoperator systems under position and rate control. In *Proceedings of IEEE/RSJ International Conference on Intelligent Robots and Systems*, Pittsburgh, PA, USA, August 1995.
343. W.H. Zhu and S.E. Salcudean. Stability guaranteed teleoperation: an adaptive motion/force control approach. *IEEE Transactions on Automatic Control*, 45(11):1951–1969, 2000.

Index

+ pairing 16, 20, 188

adjoint matrix 81
admittance
  causality 97
atlas 202

behavior 2, 17, 78
  representable by a differential equation 170
behavioral
  approach 2, 168
  deviation 170
  dynamical equations 6
  equations 3
bundle
  cotangent 204
  tangent 203
  vector 16, 188, 206

Casimir function 55
chart
  compatible 201
  coordinate 201
codistribution 18, 206
communication channel 130
  Internet-like 148
  lossless 136
  packet switching 147
  passive 132
  scattering based 135
  with variable delay 151
control as interconnection 51, 79
  energy shaping as 55
controller
  admissible 52
  family of admissible 52
covector 13, 15, 203
  field 204

damper 82
  line 140
damping injection 42
differentiable manifold 202
  dimension of 202
Dirac structure 16
  extended 188
  regular point of 20
dissipation 22
  inequality 34
  obstacle 48
dissipative system 34
distribution 18, 205
duality product 13, 15, 188
dynamical system 5
  I/O 9
  I/S/O 9, 23
  interconnection 51
  state space 7
  with latent variables 7

effort 13
  interaction 168
energy
  balancing 45
  bookkeeping 96
  leaks 89
  leap 95

lost packets 150
packet 149
shaping 44
sharing 193

fiber 16, 188, 206
flow 13
interaction 168
force
scaling 110

grasping
strategy 87

haptic interface 89
intrinsically passive 100
HLS strategy 152, 156
homogeneous
matrix 80, 209
Hunt-Crossley model 167, 172, 174

IDA-PBC 63
impedance
causality 97, 105
characteristic 26, 135
control 77
interaction 130
intrinsically passive control of 78
interconnection
and damping assignment 63
power preserving 52
power preserving 14
interpolation
passivity preserving 155
IPC 79, 131, 140, 175
digital 147
for robotic hands 86
for robotic arms 81

Jacobi-Lie brackets 205

KYP property 37

Lie algebra 208
Lie derivative 204
Lie group 80, 208
lossless
communication channel 136
system 36

map
multilinear 206
tangent 205
mathematical model 2
with latent variables 4

outcome 2
extended port 169
port 169

parameters
variation 185, 189
passive
system 36
asymptotic stabilization of 39
passivity
sampled data 98
theory 34
physical system 10
PO/PC 88, 125
port
behavior 168
interaction 78
port-Hamiltonian system 131, 187
explicit 22
extended 188
implicit 16
implicit with dissipation 17
passivity analysis of 40
sampled 91
sampled in scattering form 108
with delayed output 105
with scattering based interaction port 107
power 13
of a port outcome 169
port 13, 15, 25, 52
scaling 113
wave 29, 136
prolongation 5, 170

quantization 101

representation
kernel 20
scattering 25
RNP strategy 151, 153, 155, 178
robot
anthropomorphic 81
defective 84

robotic hand 86

saturation 66
scattering 135
  discrete time 148
  power decomposition 29
  variable 29
  wave 105
space
  cotangent 203
  dual 12, 15
  equating 3, 6, 8
  jet 6, 170
  tangent 203
spring
  spatial 82
  variable rest length 82
state
  extended space 188
  non parametric part of 187
  parametric part of 187
storage
  available 35
  function 34
strictly passive
  system 37
supervisory system 79, 131
supply rate 34

TCP/IP protocol 149
telegrasping 142
telemanipulation 106
  bilateral 129
  intrinsically passive bilateral 137
  intrinsically passive bilateral 173
tensor 207
  bundle 207
  field 207
translation
  left 208
  right 208
transparency 165, 171
  deviation index 171
twist 31, 80, 209

universe 2
  of port outcomes 170

variable
  energy 11, 15
  latent 3
  manifest 3, 17
  power conjugated 13, 135
  scattering 29
vector 13, 15
  bundle 16, 188, 206
  field 204
  tangent 203
virtual
  coupling 88
  delayed environment 104
  environment 88
  object 82
  point 79, 174

wave reflection 138
wrench 31, 80, 209

zero state
  detectability 38
  observability 38

Springer Tracts in Advanced Robotics

Edited by B. Siciliano, O. Khatib, and F. Groen

Vol. 28: Thrun, S.; Brooks, R.; Durrant-Whyte, H. (Eds.)
Robotics Research – Results of the 12th International Symposium ISRR
602 p. 2007 [978-3-540-48110-2]

Vol. 27: Montemerlo, M.; Thrun, S.
FastSLAM – A Scalable Method for the Simultaneous Localization and Mapping Problem in Robotics
120 p. 2007 [978-3-540-46399-3]

Vol. 26: Taylor, G.; Kleeman, L.
Visual Perception and Robotic Manipulation – 3D Object Recognition, Tracking and Hand-Eye Coordination
218 p. 2007 [978-3-540-33454-5]

Vol. 25: Corke, P.; Sukkarieh, S. (Eds.)
Field and Service Robotics – Results of the 5th International Conference
580 p. 2006 [978-3-540-33452-1]

Vol. 24: Yuta, S.; Asama, H.; Thrun, S.; Prassler, E.; Tsubouchi, T. (Eds.)
Field and Service Robotics – Recent Advances in Research and Applications
550 p. 2006 [978-3-540-32801-8]

Vol. 23: Andrade-Cetto, J,; Sanfeliu, A.
Environment Learning for Indoor Mobile Robots – A Stochastic State Estimation Approach to Simultaneous Localization and Map Building
130 p. 2006 [978-3-540-32795-0]

Vol. 22: Christensen, H.I. (Ed.)
European Robotics Symposium 2006
209 p. 2006 [978-3-540-32688-5]

Vol. 21: Ang Jr., H.; Khatib, O. (Eds.)
Experimental Robotics IX – The 9th International Symposium on Experimental Robotics
618 p. 2006 [978-3-540-28816-9]

Vol. 20: Xu, Y.; Ou, Y.
Control of Single Wheel Robots
188 p. 2005 [978-3-540-28184-9]

Vol. 19: Lefebvre, T.; Bruyninckx, H.; De Schutter, J.
Nonlinear Kalman Filtering for Force-Controlled Robot Tasks
280 p. 2005 [978-3-540-28023-1]

Vol. 18: Barbagli, F.; Prattichizzo, D.; Salisbury, K. (Eds.)
Multi-point Interaction with Real and Virtual Objects
281 p. 2005 [978-3-540-26036-3]

Vol. 17: Erdmann, M.; Hsu, D.; Overmars, M.; van der Stappen, F.A (Eds.)
Algorithmic Foundations of Robotics VI
472 p. 2005 [978-3-540-25728-8]

Vol. 16: Cuesta, F.; Ollero, A.
Intelligent Mobile Robot Navigation
224 p. 2005 [978-3-540-23956-7]

Vol. 15: Dario, P.; Chatila R. (Eds.)
Robotics Research – The Eleventh International Symposium
595 p. 2005 [978-3-540-23214-8]

Vol. 14: Prassler, E.; Lawitzky, G.; Stopp, A.; Grunwald, G.; Hägele, M.; Dillmann, R.; Iossifidis. I. (Eds.)
Advances in Human-Robot Interaction
414 p. 2005 [978-3-540-23211-7]

Vol. 13: Chung, W.
Nonholonomic Manipulators
115 p. 2004 [978-3-540-22108-1]

Vol. 12: Iagnemma K.; Dubowsky, S.
Mobile Robots in Rough Terrain – Estimation, Motion Planning, and Control with Application to Planetary Rovers
123 p. 2004 [978-3-540-21968-2]

Vol. 11: Kim, J.-H.; Kim, D.-H.; Kim, Y.-J.; Seow, K.-T.
Soccer Robotics
353 p. 2004 [978-3-540-21859-3]

Vol. 10: Siciliano, B.; De Luca, A.; Melchiorri, C.; Casalino, G. (Eds.)
Advances in Control of Articulated and Mobile Robots
259 p. 2004 [978-3-540-20783-2]

Vol. 9: Yamane, K.
Simulating and Generating Motions of Human Figures
176 p. 2004 [978-3-540-20317-9]

Vol. 8: Baeten, J.; De Schutter, J.
Integrated Visual Servoing and Force Control – The Task Frame Approach
198 p. 2004 [978-3-540-40475-0]

Vol. 7: Boissonnat, J.-D.; Burdick, J.; Goldberg, K.; Hutchinson, S. (Eds.)
Algorithmic Foundations of Robotics V
577 p. 2004 [978-3-540-40476-7]

Vol. 6: Jarvis, R.A.; Zelinsky, A. (Eds.)
Robotics Research – The Tenth International Symposium
580 p. 2003 [978-3-540-00550-6]

Vol. 5: Siciliano, B.; Dario, P. (Eds.)
Experimental Robotics VIil – Proceedings of the 8th International Symposium ISER02
685 p. 2003 [978-3-540-00305-2]

Vol. 4: Bicchi, A.; Christensen, H.I.; Prattichizzo, D. (Eds.)
Control Problems in Robotics
296 p. 2003 [978-3-540-00251-2]

Vol. 3: Natale, C.
Interaction Control of Robot Manipulators – Six-degrees-of-freedom Tasks
120 p. 2003 [978-3-540-00159-1]

Vol. 2: Antonelli, G.
Underwater Robots – Motion and Force Control of Vehicle-Manipulator Systems
268 p. 2006 [978-3-540-31752-4]

Vol. 1: Caccavale, F.; Villani, L. (Eds.)
Fault Diagnosis and Fault Tolerance for Mechatronic Systems – Recent Advances
191 p. 2003 [978-3-540-44159-5]

Printing: Krips bv, Meppel
Binding: Stürtz, Würzburg